XV Mexican School of Particles and Fields 2012

Journal of Physics: Conference Series Volume 468

Puebla, Mexico
6-15 September 2012

ISBN: 978-1-62993-447-1
ISSN: 1742-6588

Printed from e-media with permission by:

Curran Associates, Inc.
57 Morehouse Lane
Red Hook, NY 12571

Some format issues inherent in the e-media version may also appear in this print version.

Printed by Curran Associates, Inc. (2013)

Institute of Physics
Dirac House, Temple Back
Bristol BS1 6BE UK

Phone: 44 1 17 929 7481
Fax: 44 1 17 920 0979

techtracking@iop.org

Additional copies of this publication are available from:

Curran Associates, Inc.
57 Morehouse Lane
Red Hook, NY 12571 USA
Phone: 845-758-0400
Fax: 845-758-2634
Email: curran@proceedings.com
Web: www.proceedings.com

TABLE OF CONTENTS

List of Participants:

Name	Institution	City
Mr. ACOSTA, Alberto	Universidad de Colima	Colima
AHMAD, Aftab	IFM-UMSNH	Morelia
Mr. AHMADINIAZ, Naser	IFM-UMSNH	Morelia
ALVAREZ, Juan de Dios	IFM-UMSNH	Morelia
Ms. APARICIO MÉNDEZ, María del Rocío	FCFM-BUAP	Puebla
Dr. ARANDA-SÁNCHEZ, Jorge Isidro	FCFM-UMSNH	Morelia
ARANDA, Alfredo	Universidad de Colima	Colima
Mr. ARROYO UREÑA, Marco Antonio	FCFM-BUAP	Puebla
Dr. ARTEAGA VELAZQUEZ, Juan Carlos	IFM-UMSNH	Morelia
Mr. BEDOLLA HERNÁNDEZ, Marco Antonio	IFM-UMSNH	Morelia
Mr. BELLO MARTÍNEZ, Héctor	FCFM-BUAP	Puebla
Ms. BENITEZ GUZMÁN, Lucero Guadalupe	FCFM-BUAP	Puebla
Mr. BERRA MONTIEL, Oscar Jasel	BUAP	Puebla
BERUMEN, Francisco	Universidad Autonoma de Zacatecas	Zacatecas
Dr. BOLAÑOS, Azucena	FCFM BUAP	Puebla
BONILLA, Cesar	FCFM-BUAP	Puebla
Mr. BONILLA, Mauro Javier	INAOE	Tonantzintla
Mr. CARPINTEYRO BERNARDINO, Severiano	FCFM-BUAP	Puebla
Ms. CASTAÑEDA ALMANZA, Cosmy Polet	Universidad Autónoma de Zacatecas	Zacatecas
Mr. CASTILLA, Alejandro	DCI Universidad de Guanajuato	Norfolk
CASTILLO, Mario	FCFM-BUAP	Puebla
Ms. CASTRO MEDINA, Judith	FCFM-BUAP	Puebla
Ms. CAÑAS ORDUZ, Blanca Cecilia	CINVESTAV-IPN	México
Mr. CONDE SÁNCHEZ, José Rubén	FCFM-BUAP	Puebla
Ms. CORRAL LÓPEZ, Doribel	Universidad de Sonora	Hermosillo
Dr. CORTES MALDONADO, Ismael	CINVESTAV-IPN	MEXICO, DF
COTTI, Umberto	UMSNH	Morelia
COTZOMI, Jorge	BUAP	Puebla
Mr. CRUZ ALBARO, Eligio	FCFM-UMSNH	Morelia
Dr. CUAUTLE, Eleazar	FCFM-BUAP, ICN-UNAM	Puebla
Ms. CUEVAS ZÚÑIGA, Ana Celene	DCI Universidad de Guanajuato	León
DE ANDA, Francisco J.	Universidad de Guadalajara	Tepatitlan
De la Fuente Acosta, Eduardo	Universidad de Guadalajara	Guadalajara
Mr. DE LA FUENTE LEÓN, José Augusto	FCFM-BUAP	Puebla
Prof. DE LEÓN ACUÑA, Cederik León	UMSNH	Morelia
Prof. DELEPINE, DAVID	DCI-UGTO	leon
Ms. DIANA DEL CARMEN, Rojas Ciofalo	IFUAP	Puebla
Mr. DIAZ MENDEZ, Enrique	FCFM-BUAP	Puebla
Dr. DIAZ-CRUZ, Lorenzo	FCFM-BUAP	Puebla
Mr. DÍAZ FURLONG, Alfonso	BUAP	Puebla
Mr. DÍAZ REYNOSO, Uriel Alberto	UNAM	Puruándiro
Mr. ENCINO MARTÍNEZ, José Ramón	FCFM-UMSNH	Morelia
Mr. ESPINOSA GÓMEZ, David	FCFM-UMSNH	Morelia
Dr. FERNANDEZ TELLEZ , Arturo	FCFM-BUAP	Puebla
Ms. FÉLIX BELTRÁN, Olga Guadalupe	FCFM-BUAP	Puebla
Dr. GAITÁN LOZANO, Ricardo	FES-Cuautitlán, UNAM	Cuautitlán Izcalli

Ms. GALINDO TÉLLEZ, Aline	INAOE	Tonantzintla
Mr. GAONA, Alejandro	UAM-Cuajimalpa	Mexico City
Mr. GARCIA, Jose Didino	CINVESTAV	Mexico City
Mr. GARCÍA ARROYO, Gabriela	FCFM-UMSNH	Morelia
Mr. GONZALEZ ANAYA, Jesús Alejandro	Universidad de Guadalajara	Tepatitlan
Dr. GONZÁLEZ CANALES , Félix Francisco	FCE-BUAP	Puebla
Mr. GONZÁLEZ LÓPEZ, Juan Carlos	CINVESTAV-IPN	Naucalpan
Dr. GONZÁLEZ MÉNDEZ, Luis Xavier	INAOE	Tonantzintla
Prof. GONZÁLEZ SÁNCHEZ, Alejandro	Universidad Autónoma de Zacatecas	Zacatecas
Mr. GRADOS LUYANDO, Juan Manuel	FCFM-BUAP	Puebla
Mr. GRANADOS GONZÁLEZ, Alejandro	FCFM-BUAP	Puebla
Ms. GUITIERREZ HERRERA, Karla Cinthya	FCFM-BUAP	Puebla
Dr. GUTIÉRREZ-RODRÍGUEZ, Alejandro	Universidad Autónoma de Zacatecas	Zacatecas
Dr. GÓMEZ BOCK, Melina	BUAP	Puebla, Pue.
Mr. HERNÁNDEZ ORTIZ, Saúl Fernando	IFM-UMSNH	Morelia
Mr. HERNÁNDEZ ROSAS, Luis Alberto	ICN-UNAM	México
Ms. HERNÁNDEZ RUIZ, Maria de los Ángeles	IF-UASLP	San Luis Potosí
HERNÁNDEZ-TOMÉ, Gerardo	FCFM-BUAP	Puebla
Mr. JAVIER MORALES, Francisco	FCFM-BUAP	Puebla
Mr. JIMÉNEZ ANDRADE, Eric Faustino	FCFM-BUAP	Puebla
Mr. JIMÉNEZ MONTERO, Luis Enrique	FCFM-BUAP	Puebla
LOAIZA-BRITO, Oscar	DCI Universidad de Guanajuato	Leon
Mr. LOZANO TORRES, José Agustín	FFIA- Universidad Veracruzana	Xalapa
Mr. LÓPEZ SÁNCHEZ, Christian Mauricio	DCI Universidad de Guanajuato	LEÓN
Mr. LÓPEZ VALDOVINOS, Marlon Gerardo	FCFM-UMSNH	Morelia
Mr. LÓPEZ, Jesús	UNAM	México df.
Mr. MALDONADO SÁNCHEZ, Raúl Felipe	FCFM-BUAP	Puebla
Ms. MALDONADO, Ivonne	ICN-UNAM	México
Dr. MARTÍNEZ BRAVO, Oscar	FCFM-BUAP	Puebla
Dr. MARTINEZ HERNANDEZ, Mario Ivan	FCFM - BUAP	Puebla
Mr. MARTÍNEZ, Humberto	CINVESTAV	D.F.
Mr. MEDINA, Luis Eduardo	DCI Universidad de Guanajuato	LEÓN
Mr. MEJÍA, Jhovanny	CINVESTAV-IPN	México
Dr. MIZHER, Ana Julia	UNAM	Mexico D.F.
Dr. MONTANO ZETINA, Luis Manuel	Cinvestav	Mexico
Mr. MONTAÑO PERAZA, Javier	Universidad de Sonora	Hermosillo
Dr. MONTES DE OCA YEMHA, Jose Halim	FESC-UNAM	Mexico
Ms. MORALES REYES, Alma Roberta	FCFM-BUAP	Puebla
Dr. MORENO, Eduardo	FCFM BUAP	Puebla
Ms. MORENO RODRÍGUEZ, Lilian A.	FCFM-BUAP	Puebla
Mr. MOYOTL-ACUAHUITL, Agustin	IF-BUAP	Puebla
Prof. NAPSUCIALE, Mauro	DCI Universidad de Guanajuato	León
NAVARRETE, Efren	FESC-UNAM	Cuautitlan Izcalli
Mr. ORDUZ-DUCUARA, Javier-Andres	BUAP	Puebla
Ms. ORTEGA VIDALS, Paula	FCFM-BUAP	Puebla
Mr. PATIÑO VILLAGÓMEZ, Daniel A.	FCFM-BUAP	Puebla
PEDRAZA MORALES, maria isabel	FCFM-BUAP	Geneva
Dr. PELAYO, Rodrigo	FCFM-BUAP	Puebla
Mr. PEREZ , Carlos	Depto. Matemáticas, CINVESTAV	Mexico DF
Dr. PEREZ-LORENZANA, Abdel	CINVESTAV	Mexico, D.F.
Mr. PÉREZ ALANIS, Carlos Arturo	FCFM-UMSNH	Morelia
Ms. PÉREZ MARTÍNEZ, Adriana	FCFM-BUAP	Puebla

Mr. PÉREZ VILCHIS, Jaime	FCE-BUAP	Puebla
Mr. PONCE , Epifanio	FCFM-BUAP	Puebla
QUINTERO POVEDA, Néstor	CINVESTAV	D.F.
RAMIREZ, Monica F.	Universidad de Guadalajara	Guadalajara
RAMÍREZ QUEZADA, Maura Elizabeth	Universidad de Sonora	Hermosillo
Mr. RAMÍREZ RODRÍGUEZ, Héctor Ariel	FFIA-Universidad Veracruzana	Xalapa
Dr. RAMÍREZ ZAVALETA, Fernando	FCFM-UMSNH	Morelia
Dr. RAYA, Alfredo	IFM-UMSNH	Morelia
RAYA, Khépani	FCFM	Morelia
Mr. REYES MELÉNDEZ, Jorge	INAOE	Tonantzintla
Ms. RIVERA AGUDELO, Diana Carolina	CINVESTAV-IPN	México
Mr. RIVERA, David	FCFM-UMSNH	morelia
Mr. RODRIGUEZ CAHUANTZI, MARIO	FCFM - BUAP	Puebla
Dr. ROJAS PACHECO, Alma Dolores	Universidad de Colima	Colima
Mr. ROSADO NAVARRO, Sebastián	FCFM-BUAP	Puebla
Prof. ROSADO SÁNCHEZ, Alfonso	IF-BUAP	Puebla
Dr. SALAZAR, Humberto	FCFM-BUAP	Puebla, Puebla
Mr. SALDAÑA SALAZAR, Ulises Jesús	Instituto de Fisica (UNAM)	Mexico City
Mr. SALDAÑA ULLOA, Diego	Universidad de las Américas Puebla	Puebla
Mr. SERNA, Cesar	DCI Universidad de Guanajuato	Leon
Ms. SIES RAMOS, Laura	FCFM-BUAP	Puebla
Mr. SOSA CHÁVEZ, Arturo	FCFM-UMSNH	Morelia
Ms. SOSA SANCHEZ, Citlalli Teresa	BUAP	Puebla
STEERENBERG, Rende	CERN, Geneva, Switzerland	Geneve
Dr. TAVARES VELASCO, Gilberto	FCFM-BUAP	Puebla
Mr. THAKUR, Ravindra	IFM-UMSNH	Morelia
Dr. TORRES , Ibrahim	INAOE	Tonantzintla
Mr. TOSTADO ROBLEDO, Sergio Lennin	CINVESTAV-IPN	México
Prof. TOSCANO, Jesús	FCFM-BUAP	Puebla
Mr. TÉLLEZ LORENZO, Pascual	FCFM-BUAP	Puebla
Mr. VALDEZ TORIJA, Eduardo	FCFM-BUAP	Puebla
Mr. VARELA, Enrique	FCFM-BUAP	Puebla
Ms. VARGAS HERNÁNDEZ, Laura Gabriela	FCFM-BUAP	Puebla
Mr. VILLANUEVA RÍOS, Leonel	DCI Universidad de Guanajuato	LEÓN
Ms. XOXOCOTZI, Reyna	FCFM-BUAP	Puebla

Speakers

Matthias Neubert	Mainz, Germany
Andreas Haungs	KIT, Karlsruhe
José Ramón Espinoza	ICREA, Barcelona
Lian Tao Wang	University of Chicago
Rende Steerenberg	BE/OP
John Ellis	King's College, London
Emilio Meschi	CERN
Tiziano Camporesi	CERN
Luis Flores Castillo	U. Wisconsin
Arturo Fernández	FCFM-BUAP
Alberto Guijosa	ICN-UNAM
Alberto Carramiñana	INAOE
Julián Félix	U. de Guanajuato
Abdel Pérez Lorenzana	Cinvestav
Alfredo Aranda	U. de Colima
Alexis Aguilar	ICN-UNAM

Journal of Physics: Conference Series **468** (2013) 012001 doi:10.1088/1742-6596/468/1/012001

The $Z' \to \tau\mu$ decay in the sequential Z model

J I Aranda[(a)], **J Montaño**[(b)], **F Ramírez-Zavaleta**[(a)], **J J Toscano**[(c)] **and E S Tututi**[(a)]

[(a)]Facultad de Ciencias Físico Matemáticas, Universidad Michoacana de San Nicolás de Hidalgo, Avenida Francisco J. Mújica S/N, 58060, Morelia, Michoacán, México.
[(b)]Departamento de Física, CINVESTAV, Apartado Postal 14-740, 07000, México, D. F., México.
[(c)]Facultad de Ciencias Físico Matemáticas, Benemérita Universidad Autónoma de Puebla, Apartado Postal 1152, Puebla, Puebla, México.

E-mail: feramirez@umich.mx

Abstract. A calculation of the strength of the lepton flavor violating $Z'\mu\tau$ coupling is presented by using the most general renormalizable Lagrangian that includes lepton flavor violation. We employ the experimental restriction over the $\tau \to \mu\mu^+\mu^-$ decay in the context of the sequential Z model to bound the mentioned coupling. It was found that the most restrictive result is provided by a vector-like coupling, denoted as $|\Omega_{\mu\tau}|^2$, finding around 10^{-2} for a Z' boson mass of 2 TeV. We used this information to estimate the branching ratio for the $Z' \to \tau\mu$ decay. According to the analyzed model the associated branching ratio is of the order of 10^{-2} for a Z' boson mass around 2 TeV.

1. Introduction

The presence of a new neutral massive gauge boson (Z') is predicted in the context of numerous extensions of the Standard Model (SM). The $SU_C(3) \times SU_L(2) \times U_Y(1) \times U'(1)$ extended electroweak gauge group is the simplest model that predicts the existence of the Z' boson [1]. Although absent in the SM, the phenomena of lepton flavor violation can arise in many of its well-motived extensions. One interesting feature of most models beyond the SM is the presence of generalized Yukawa sectors, which favor nondiagonal transitions mediated by neutral massive spin-1 particles [1]. Currently, the experimental data have evidenced the existence of neutrino oscillations, which in simple words, tell us that lepton flavor conservation is violated in nature. However, the only signal of lepton flavor violation comes from transitions between neutral leptons. Lepton flavor violating transitions between charged leptons constitute an interesting subject to study, since if they occur in nature it represents further evidence of lepton flavor violation.

The paper is organized as follows. In Section II a discussion of the extended model is presented. Section III is devoted to present the strength of the $Z'\mu\tau$ coupling. In Section IV, we discuss our results for the $Z' \to \tau\mu$ decay. Finally, in Section V the conclusions are presented.

2. The extended model

We consider the most general renormalizable Lagrangian, which includes the lepton flavor violation mediated by a new neutral massive gauge boson, coming from any extended or grand unification model [1, 2], which can be written as

$$\mathcal{L}_{NC} = \sum_{i,j} \Big[\bar{f}_i \gamma^\alpha (\Omega_{Lf_if_j} P_L + \Omega_{Rf_if_j} P_R) f_j + \bar{f}_j \gamma^\alpha (\Omega^*_{Lf_jf_i} P_L + \Omega^*_{Rf_jf_i} P_R) f_i \Big] Z'_\alpha,$$

where f_i is any fermion of the SM, $P_{L,R} = \frac{1}{2}(1 \pm \gamma_5)$ are the chiral projectors, and Z'_α is a new neutral massive gauge boson predicted by several extensions of the SM. The $\Omega_{Ll_il_j}$, $\Omega_{Rl_il_j}$ parameters represent the strength of the $Z'l_il_j$ coupling, where l_i is any charged lepton of the SM. In the rest of this work we will assume that $\Omega_{Ll_il_j} = \Omega_{Ll_jl_i}$ and $\Omega_{Rl_il_j} = \Omega_{Rl_jl_i}$. The above Lagrangian includes both flavor-conserving and flavor-violating couplings mediated by a Z' gauge boson.

This work is oriented to study the impact of lepton flavor-violating couplings mediated by a Z' boson in the $Z' \to \tau\mu$ decay. For this purpose we need to bound the lepton flavor-violating coupling $Z'\tau\mu$. This task will be realized by using the experimental restriction for the lepton flavor-violating $\tau \to \mu\mu^+\mu^-$ decay, since it is proven that this process offers the more restrictive bounds [2]. Moreover, in order to calculate the branching ratio of the $Z' \to \tau\mu$ process it is necessary to know the Z' total width decay, which mainly depends on the $Z' \to f_i\bar{f}_i$ decays. These flavor-conserving couplings are model dependent. The flavor-conserving couplings, $Q^{f_i}_{L,R}$ [2], are related to the Ω couplings as $\Omega_{Lf_if_i} = -g_2\, Q^{f_i}_L$ and $\Omega_{Rf_if_i} = -g_2\, Q^{f_i}_R$, where $g_2 = g/\cos\theta_W$ is the gauge coupling of the Z' boson.

3. Calculation of the $Z'\mu\tau$ coupling

The contribution of the flavor-violating $Z'\tau\mu$ vertex to the $\tau \to \mu\mu^+\mu^-$ decay is given through the Feynman diagram shown in Fig. 1. Since the three-body decay of the tau lepton comes from a tree-level Feynman diagram we only present their corresponding branching ratio

$$Br(\tau \to \mu\mu^+\mu^-) = \frac{g_2^2}{384\pi^3}\Big(h_1(m_{Z'})(|Q^e_L\Omega_{L\mu\tau}|^2 + |Q^e_R\Omega_{R\mu\tau}|^2) + h_2(m_{Z'})(|Q^e_L\Omega_{R\mu\tau}|^2 + |Q^e_R\Omega_{L\mu\tau}|^2)\Big)\frac{m_\tau}{\Gamma_\tau},$$

where

$$h_1(m_{Z'}) = \int_0^1 dx \frac{2x-1}{(x-1+m^2_{Z'}/m^2_\tau)}(2(7-4x)x-5),$$

$$h_2(m_{Z'}) = \int_0^1 dx \frac{2x-1}{(x-1+m^2_{Z'}/m^2_\tau)}(1-2(x-1)x).$$

The branching ratio of the $\tau \to \mu\mu^+\mu^-$ decay must be less than the corresponding experimental restriction, $\mathrm{Br}_{\mathrm{Exp}}(\tau \to \mu\mu^+\mu^-) < 2.1 \times 10^{-8}$ [3], which allow us to get a bound on the flavor-violating $|\Omega_{L\mu\tau}|^2$ and $|\Omega_{R\mu\tau}|^2$ parameters.

For a Z' boson mass ranging in the interval 1 TeV $< m_{Z'} <$ 2 TeV, the $h_2(m_{Z'})$ function is suppressed with respect to the $h_1(m_{Z'})$ function about six orders of magnitude. Thus, the interference term, $h_2(m_{Z'})(|Q^e_L\Omega_{R\mu\tau}|^2 + |Q^e_R\Omega_{L\mu\tau}|^2)$, can be neglected, which simplifies the calculation of $|\Omega_{L\mu\tau}|^2$ and $|\Omega_{R\mu\tau}|^2$ parameters. Therefore, the branching ratio becomes

$$Br(\tau \to \mu\mu^+\mu^-) = \frac{g_2^2}{384\pi^3} h_1(m_{Z'})(|Q^e_L\Omega_{L\mu\tau}|^2 + |Q^e_R\Omega_{R\mu\tau}|^2)\frac{m_\tau}{\Gamma_\tau}. \quad (1)$$

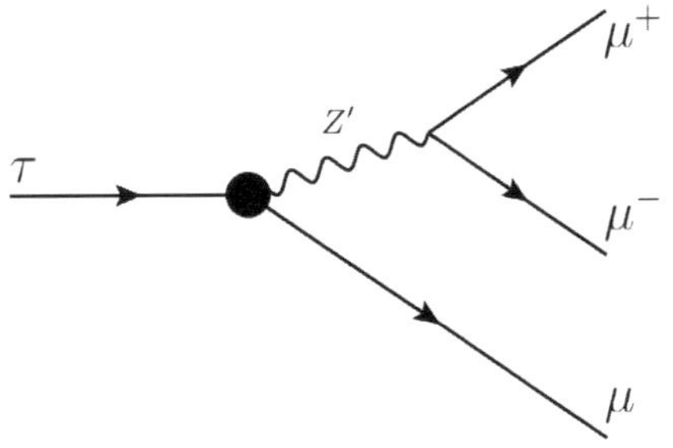

Figure 1. Feynman diagram contributing to the $\tau \to \mu\mu^+\mu^-$ decay.

If we consider a vector-like coupling implies that $\Omega_{L\mu\tau} = \Omega_{R\mu\tau}$, for simplicity, we define $\Omega_{L\mu\tau} \equiv \Omega_{\mu\tau}$. After solving for the $|\Omega_{\mu\tau}|^2$ parameter in Eq. (1) we reach the following inequality

$$|\Omega_{\mu\tau}|^2 < 384\,\pi^3 \frac{\Gamma_\tau}{m_\tau} \frac{\mathrm{Br}_{\mathrm{Exp}}(\tau \to \mu\mu^+\mu^-)}{T_2\, h_1(m_{Z'})}, \tag{2}$$

where $T_2 = g_2^2({Q_L^e}^2 + {Q_R^e}^2)$.

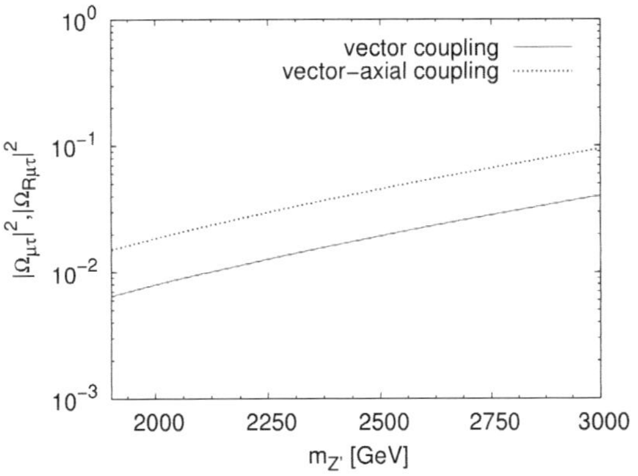

Figure 2. Behavior of $|\Omega_{\mu\tau}|^2$, $|\Omega_{R\mu\tau}|^2$ as a function of the Z' boson mass.

In Fig. 2 it can be appreciated the maximum of the $|\Omega_{\mu\tau}|^2$ parameter derived from Eq. (2). We note that its behavior is monotone increasing. The strength of the $|\Omega_{\mu\tau}|^2$ parameter varies from 10^{-3} to 10^{-2} for a Z' boson mass within the range 1.9 TeV $< m_{Z'} <$ 3 TeV.

In considering a vector-axial coupling, we take $\Omega_{L\mu\tau} = 0$. After solving for the $|\Omega_{R\mu\tau}|^2$ parameter in Eq. (1) we reach the following inequality

$$|\Omega_{R\mu\tau}|^2 < 384\,\pi^3 \frac{\Gamma_\tau}{m_\tau} \frac{\mathrm{Br}_{\mathrm{Exp}}(\tau \to \mu\mu^+\mu^-)}{T_4\, h_1(m_{Z'})}, \tag{3}$$

where $T_4 = g_2^2 {Q_R^e}^2$. In Fig. 2 we show the maximum of the $|\Omega_{R\mu\tau}|^2$ parameter derived from the inequality in Eq. (3). We note that its behavior is monotone increasing. The strength of the $|\Omega_{R\mu\tau}|^2$ parameter is of the order of 10^{-2} for a Z' boson mass within the range 1.9 TeV $< m_{Z'} <$ 3 TeV.

4. The $Z' \to \tau\mu$ decay

We resort to the bounds obtained from the previous analysis. As discussed above, we are interested in studying the $Z' \to \tau\mu$ decay, whose branching ratio can be written as

$$\mathrm{Br}(Z' \to \tau\mu) = \frac{m_{Z'}}{24\pi\Gamma_{Z'}}\left[\left(1 - \frac{(m_\tau + m_\mu)^2}{m_{Z'}^2}\right)\left(1 - \frac{(m_\tau - m_\mu)^2}{m_{Z'}^2}\right)\right]^{1/2}$$
$$\times\left[\left(2 - \frac{m_\tau^2 + m_\mu^2}{m_{Z'}^2} - \frac{(m_\tau^2 - m_\mu^2)^2}{m_{Z'}^4}\right)(|\Omega_{L\mu\tau}|^2 + |\Omega_{R\mu\tau}|^2)\right].$$

where $\Gamma_{Z'}$ is the Z' total decay width. On account $\Gamma_{Z'}$ we include the total possible flavor-conserving and flavor-violating decay modes [2, 4], namely $\nu_e\bar{\nu}_e$, $\nu_\mu\bar{\nu}_\mu$, $\nu_\tau\bar{\nu}_\tau$, $e\bar{e}$, $\mu\bar{\mu}$, $\tau\bar{\tau}$, $u\bar{u}$, $c\bar{c}$, $t\bar{t}$, $d\bar{d}$, $s\bar{s}$, $b\bar{b}$, $\bar{u}c + u\bar{c}$, $\bar{t}c + t\bar{c}$, and $\bar{\tau}\mu + \tau\bar{\mu}$.

In Fig. 3 we show the branching ratio for the $Z' \to \tau\mu$ decay in the context of the sequential Z model. From this figure, it is more feasible to observe vector-axial lepton flavor violation mediated by a Z' gauge boson, whose strength can be as higher as 8×10^{-2}.

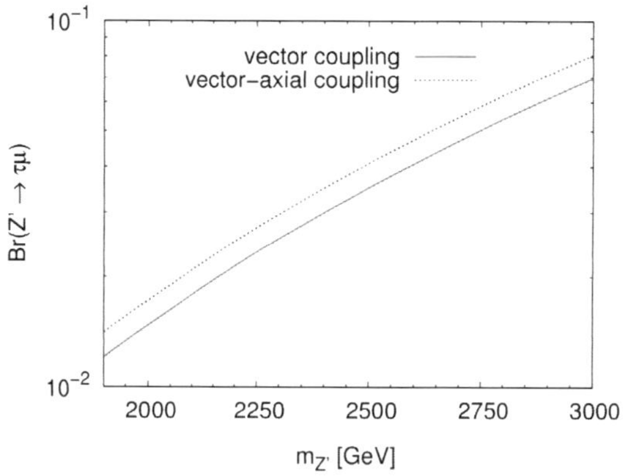

Figure 3. The branching ratio for the $Z' \to \tau\mu$ decay as function of the Z' boson mass.

5. Conclusions

Our prediction for $\mathrm{Br}(Z' \to \tau\mu)$ is of the order of 10^{-2}, in a broad range of the Z' gauge boson mass studied. This bound enables the possibility of lepton flavor violation being observed by Z' mediated processes. Our results show that current experimental constraint on the lepton flavor violating $\tau \to \mu\mu^+\mu^-$ decay restricts weakly this phenomenon.

Acknowledgments

This work has been partially supported by CIC-UMSNH and SNI (México).

References

[1] P. Langacker, Rev. Mod. Phys. **81**, 1199 (2008); L. S. Durkin and P. Langacker, Phys. Lett. B **166**, 436 (1986); M. Cvetic and P. Langacker, Proceedings of Ottawa 1992: Beyond the standard model 3, 454-458, (1992).

[2] J. I. Aranda, J. Montaño, F. Ramírez-Zavaleta, J. J. Toscano, and E. S. Tututi, Phys. Rev. D **86**, 035008 (2012).

[3] K. Nakamura *et. al.*, J. Phys. G: Nucl. Part. Phys. **37** 075021, (2010).

[4] A. Arhrib *et al.*, Phys. Rev. D **73**, 075015 (2006); J. I. Aranda, *et. al.* J. Phys. G: Nucl. Part. Phys. **38**, 045006 (2011).

Journal of Physics: Conference Series **468** (2013) 012002 doi:10.1088/1742-6596/468/1/012002

New physics effects of a fifth dimension on the $t \to cg$ decay

J I Aranda[(a)], **M G López-Valdovinos**[(a)], **F Ramírez-Zavaleta**[(a)], **J J Toscano**[(b)] **and E S Tututi**[(a)]

[(a)]Facultad de Ciencias Físico Matemáticas, Universidad Michoacana de San Nicolás de Hidalgo, Avenida Francisco J. Mújica S/N, 58060, Morelia, Michoacán, México.
[(b)]Facultad de Ciencias Físico Matemáticas, Benemérita Universidad Autónoma de Puebla, Apartado Postal 1152, Puebla, Puebla, México.

E-mail: feramirez@umich.mx

Abstract. The top quark is the heaviest elementary particle known, with a mass close to electroweak symmetry breaking energy scale. Because of this, its dynamics can be very sensitive to new physics effects. In this work we study the one-loop effects of a fifth spatial dimension on the strong couplings for the rare $t \to cg$ decay. The analysis is done in the context of the standard model in five dimensions, with the fifth coordinate compactificated on the orbifold S^1/Z_2 of radius R.

1. Introduction

The top quark (t) is the heaviest particle predicted by the Standard Model (SM). It is so heavy that may provide a link with new physics beyond the Fermi scale ($v = 246$ GeV). An obvious scenario where new physics effects may be discerned is through their couplings with the gauge bosons of the SM, such as their one-loop level couplings with SM neutral gauge bosons, namely the photon, the weak boson Z and gluons of the strong force g. This work proposes to study the impact of a compactificated fifth dimension on the vertex function associated with the tcg coupling to one-loop level.

The phenomenological implications of extra dimensions on SM observables have been the subject of considerable interest in the literature since Antoniadis, Arkani-Hamed, Dimopoulos and Dvali [1] argued that relatively large dimensions could be detected at the scale of TeVs. In most scenarios proposed, our three-dimensional space is a 3-brane embedded in a D-dimensional spacetime, which is known as the bulk. Of course, if there are additional dimensions, these must be smaller than the smallest distances explored. In our case, we assume that the fifth dimension is compactificated on the S^1/Z_2 orbifold of radius R. As a result of the compactification, the fields propagating in the bulk can be expressed in a number of states known as tower of Kaluza-Klein (KK), with individual KK excitations labeled by number modes. SM fields correspond to the zero mode.

The paper is organized as follows. In Section II a discussion of the standard model in five dimensions is presented. In Section III, we analyze our results for the $t \to cg$ decay. Finally, in Section IV the conclusions are presented.

2. The standard model in five dimensions

We consider the Standard Model defined in a five-dimensional flat space-time, in which the fifth coordinate is compactified as a circle of radius R. The four-dimensional coordinates will be denoted by x, whereas the fifth-dimension coordinate will be represented by y. For an universal extra dimension, one assumes that all fields and gauge parameters are periodic functions on this coordinate and expands them in Fourier series along it. In general, for a given field or gauge parameter, one has:

$$\varphi(x,y) = \frac{1}{\sqrt{2\pi R}}\varphi^{(0)}(x) + \sum_{n=1}^{\infty}\left[\varphi^{(n)+}(x)\cos\left(\frac{ny}{R}\right) + \varphi^{(n)-}(x)\sin\left(\frac{ny}{R}\right)\right],$$

where the zero mode $\varphi^{(0)}(x)$ is identified as the corresponding four dimensional SM field and the $\varphi^{(n)\pm}$ ones are recognized as its KK excitations [2, 3].

In the context of the Standard Model gauge group in five dimensions, $[\mathrm{SU_C}(3) \times \mathrm{SU_L}(2) \times \mathrm{U_Y}(1)]_5$, the effective action can be written as

$$S_{\text{eff}} = \int d^4x\, \mathcal{L}^{\text{eff}}_{4\text{SM}},$$

with $\mathcal{L}^{\text{eff}}_{4\text{SM}} = \int_0^{2\pi R} dy \left[\mathcal{L}_{5\text{SM}} + \sum_N^{\infty} \frac{\beta_N g_5^{N_1}}{M_s^{N_2}} \mathcal{O}_N\right]$ [2, 3]. In the above expression, $\mathcal{L}_{5\text{SM}}$ is the five-dimensional version of the SM, whose structure in terms of KK modes will be discussed with some detail below. The $\mathcal{O}_N$ are operators of canonical dimension $N > 5$, M_s is the energy scale above the new physics manifests directly, and β_N is a dimensionless parameter that depends on the details of the underlying physics. In the above Lagrangian, it is assumed that all the independent operators that respect the five-dimensional Lorentz and gauge symmetries are included and each of them is multiplied by an unknown dimensionless parameter β_N. The canonical dimension of each term of the series is appropriately corrected by introducing factors containing powers of the dimensionful coupling constants g_5 (the rule is to introduce a g_5 per each curvature appearing in the $\mathcal{O}_N$-invariant) and the M_s scale.

The corresponding four-dimensional Lagragian can be expressed as

$$\mathcal{L}_{4\text{SM}} = \int_0^{2\pi R} dy\, (\mathcal{L}_{5\text{YM}} + \mathcal{L}_{5\text{H}} + \mathcal{L}_{5\text{C}} + \mathcal{L}_{5\text{Y}}),$$

where $\mathcal{L}_{5\text{YM}}$, $\mathcal{L}_{5\text{H}}$, $\mathcal{L}_{5\text{C}}$, and $\mathcal{L}_{5\text{Y}}$ stand for the five-dimensional Yang-Mills, Higgs, currents, and Yukawa sectors, respectively.

The interest of this work is to explore the impact of KK excited modes of the SM W boson and the KK excitations of charged physical scalars ($H^{(n)\pm}$) on the $t \to cg$ decay. The contribution of KK excited modes will be compared with the theoretical results obtained in the Standard Model. To this end, it is necessary to know the physical vertices of the theory which induce this process. Since this work corresponds to a preliminary study which only analyzes the contributions of KK physical charged scalars modes, we present the Lagrangians containing the $q^{(0)}q^{(n)}H^{(n)\pm}$ and $g^{(0)}q^{(n)}q^{(n)}$ couplings, where $q^{(0)}$ represents the SM quarks, $q^{(n)}$ are KK quarks modes and $g^{(0)}$ symbolizes the SM gluon.

The couplings of scalars with pairs of quarks arise from the Yukawa sector. The Lagrangian containing $q^{(0)}q^{(n)}H^{(n)\pm}$ couplings is given by [3]

$$\begin{aligned}\mathcal{L}_{q^{(0)}q^{(n)}H^{(n)\pm}} = & \frac{g c_\alpha}{\sqrt{2} m_{W^{(0)}}} H^{(n)+} \left[\bar{U}^{(0)} P_R V_d^{(0)} \left(\sin\frac{\alpha_d^{(n)}}{2} D^{(n)} + \cos\frac{\alpha_d^{(n)}}{2} \hat{D}^{(n)} \right) \right. \\ & \left. + \left(\cos\frac{\alpha_u^{(n)}}{2} \bar{U}^{(n)} + \sin\frac{\alpha_u^{(n)}}{2} \bar{\hat{U}}^{(n)} \right) P_R V_d^{(0)} D^{(0)} \right] + \text{H.c.} \\ & - \frac{g c_\alpha}{\sqrt{2} m_W} H^{(n)-} \left[\bar{D}^{(0)} P_R V_u^{(0)\dagger} \left(\sin\frac{\alpha_u^{(n)}}{2} U^{(n)} + \cos\frac{\alpha_u^{(n)}}{2} \hat{U}^{(n)} \right) \right. \\ & \left. + \left(\cos\frac{\alpha_d^{(n)}}{2} \bar{D}^{(n)} + \sin\frac{\alpha_d^{(n)}}{2} \bar{\hat{D}}^{(n)} \right) P_R V_u^{(0)\dagger} U^{(0)} \right] + \text{H.c.},\end{aligned}$$

where $V_d^{(0)} = K M_d^{(0)}$ and $V_u^{(0)} = K M_u^{(0)}$, with $K = V_L^u V_L^{d\dagger}$ being the Cabibbo-Kobayashi-Maskawa matrix and $M_u^{(0)}$ ($M_d^{(0)}$) is the diagonal mass matrix of the SM quarks of type up (down).

The couplings of quarks to SM gluons are

$$\mathcal{L}_{g^{(0)}q^{(n)}q^{(n)}} = g_s \sum_{q=u,d,\ldots} \left(\bar{q}^{(0)} \gamma^\mu \frac{\lambda^a}{2} q^{(0)} + \bar{q}^{(n)} \gamma^\mu \frac{\lambda^a}{2} q^{(n)} + \bar{\hat{q}}^{(n)} \gamma^\mu \frac{\lambda^a}{2} \hat{q}^{(n)} \right) g_\mu^{(0)a}.$$

3. The $t \to cg$ decay

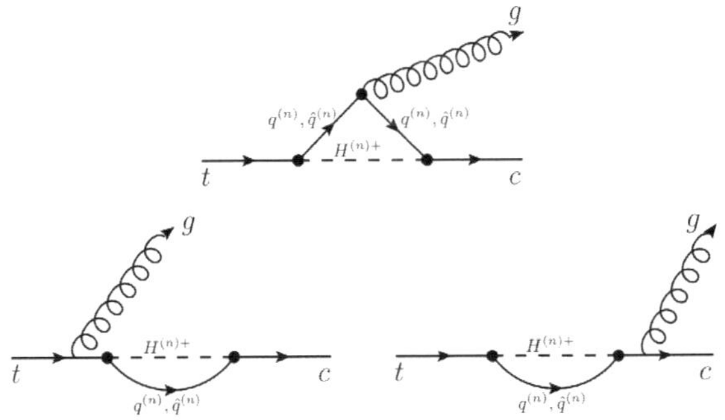

Figure 1. Feynman diagrams contributing to the $t \to cg$ decay.

The contribution of the charged scalar $H^{(n)+}$ to the $t \to cg$ decay is calculated using the Feynman diagrams shown in Figure 1. The one-loop amplitude of the $t \to cg$ process for arbitrary internal down quark is evaluated via the Passarino-Veltman reduction scheme and can be reduced to

$$\begin{aligned}\mathcal{M} = & \sum_{n=1}^{\infty} \sum_{d=d,s,b} \frac{i\sqrt{\alpha_s}\,\alpha\, K_{td}\, K_{cd}\, m_d^2}{16\sqrt{\pi}\sin^2\theta_W\, m_W^2\, m_t} \left(\frac{n^2}{n^2 + m_W^2 R^2} \right) \\ & \times \bar{u}(p_c) F_H^{(n)} \lambda^a\, P_L\, \sigma^{\mu\beta} q_\beta\, \epsilon_\mu^{a*}(q)\, u(p_t),\end{aligned}$$

where

$$\begin{aligned}F_H^{(n)} = & 1 + \frac{m_{H^{(n)+}}^2 - m_{d^{(n)}}^2}{m_t^2} \left(B_0(0, m_{d^{(n)}}^2, m_{H^{(n)+}}^2) - B_0(m_t^2, m_{d^{(n)}}^2, m_{H^{(n)+}}^2) \right) \\ & + 2 m_{d^{(n)}}^2\, C_0(m_t^2, 0, 0, m_{d^{(n)}}^2, m_{H^{(n)+}}^2, m_{d^{(n)}}^2),\end{aligned}$$

and $m^2_{H^{(n)+}} = \frac{n^2}{R^2} + m^2_W$, $m^2_{d^{(n)}} = \frac{n^2}{R^2} + m^2_d$.

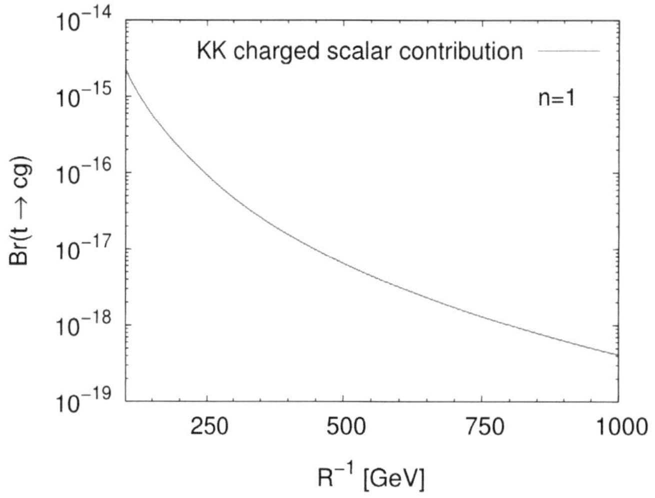

Figure 2. Behavior of $\mathrm{Br}(t \to cg)$ as a function of the R^{-1} for $n = 1$.

Notice that only a magnetic dipole contribution is present, which is ultraviolet finite. After squaring $\mathcal{M}$, one obtains the associated branching ratio by means of the formula

$$\mathrm{Br}(t \to cg) = \frac{|\mathcal{M}|^2}{16\,\pi\,m_t\,\Gamma_t}.$$

where Γ_t is the top quark total decay width.

The behavior of $\mathrm{Br}(t \to cg)$ can be appreciated in Figure 2. It can be seen from this figure that the contribution of charged physical scalars to the branching ratio of the $t \to cg$ decay is very suppressed, which varies between 10^{-15} to 10^{-19} for a R^{-1} in the range of 200 GeV $< R^{-1} <$ 1 TeV. If we compare our result with the prediction obtained in the standard model is clear that the physical charged scalar contribution would be unobservable, since the SM branching ratio of the $t \to cg$ process is of the order of 10^{-11} [4].

4. Conclusions

It is shown that the new physics effects decouple in the limit of a small compactification scale R. The contribution of the excited KK charged scalar modes to $t \to cg$ process is suppressed with respect to the SM radiative correction by up to 8 orders of magnitude.

Acknowledgments

This work has been partially supported by CIC-UMSNH and SNI (México).

References

[1] I. Antoniadis, Phys. Lett. **B** 246, 377 (1990); N. Arkani-Hamed, S. Dimopoulos, and G. R. Dvali, Phys. Lett. **B** 429, 263 (1998); I. Antoniadis, N. Arkani-Hamed, S. Dimopoulos, and G. R. Dvali, Phys. Lett. **B** 436, 257 (1998).
[2] H. Novales-Sánchez and J. J. Toscano, Phys. Rev. D **82**, 116012 (2010); H. Novales-Sánchez and J. J. Toscano, Phys. Rev. D **84**, 076010 (2011).
[3] A. Cordero-Cid, M. Gómez-Bock, H. Novales-Sánchez and J. J. Toscano, arXiv:1108.2926 [hep-ph].
[4] G. Eilam, J. L. Hewett, and A. Soni, Phys. Rev. D **44**, 1473 (1991).

Journal of Physics: Conference Series **468** (2013) 012003 doi:10.1088/1742-6596/468/1/012003

Lepton Asymmetry with Primordial Magnetic Fields

Roberto Cota[1], Juan Barranco[1], David Delepine[1] and Shaaban Khalil[2,3]

[1] Division de Ciencias e Ingenierías, Universidad de Guanajuato, Campus Leon, C.P. 37150, León, Guanajuato, México.

[2] Centre for Theoretical Physics, Zewail City of Science and Technology, Sheikh Zayed, 12588, Giza, Egypt.

[3] Department of Mathematics, Faculty of Science, Ain Shams University, Cairo, Egypt.

E-mail: cota@fisica.ugto.mx, jbarranc@fisica.ugto.mx, delepine@fisica.ugto.mx, skhalil@zewailcity.edu.eg

Abstract. If the neutrino are Majorana particles, the neutrino spin flavor precession induced $\Delta L = 2$ processes which could be relevant for leptogenesis depending on the strength of neutrino magnetic moments and magnetic fields. Although the extra galactic magnetic fields is extremely weak at present time (about 10^{-9} Gauss), the primordial magnetic field at the electroweak scale could be strong (of order 10^{17} Gauss). Therefore, at this scale, the effects of the spin flavor precession could not be negligible. Using present limit on neutrino magnetic moments, we show that the lepton asymmetry may be reduced by 50% due to the spin flavor precession induced by strong primordial magnetic fields. In addition, the leptogenesis will have different feature from the standard scenario of leptogenesis, where the lepton asymmetry continues to oscillate even after the electroweak phase transition.

1. Introduction

Observations indicate that the baryonic matter we are made off is the remanent of a small matter-antimatter asymmetry originated at the early universe[1, 2]. Leptogenesis is a scenario where it is possible to connect the smallness of the light Majorana neutrino masses with a mechanism to generate this matter-antimatter asymmetry. This mechanism requires right-handed Majorana neutrinos that decay out-of-equilibrium. This decay process, combined with non-perturbative anomalous electro-weak processes, can generate the baryon number in the universe [3, 4, 5, 6, 7]. In this model of baryogenesis, it is expected that the lepton asymmetry to be of the same order of magnitude that of the baryon asymmetry, due to sphaleron effects that are relevant for temperature from 10^{12} GeV to 100 GeV [8, 9]. The measurement of the Baryon Asymmetry of the Universe (BAU) through the anistropies of the cosmic microwave background radiation (CMB) together with other cosmological observations at a very high level of precision have strongly constrained BAU, that is parameterized by the ratio of baryon number to photon number: $\eta_B = N_B/N_\gamma$. Recent analysis [10] implies that

$$\eta_B = (5.8 \pm 0.27) \times 10^{-10}, \quad (1)$$

which show that the measurement of baryon asymmetry is achieved with an error less than 5%.

Unfortunately, the lepton asymmetry is not precisely measured as the baron asymmetry. Recently, an attempt to constraint the lepton asymmetry from WMAP and nucleosynthesis has been done [11]. The following limits on $\eta_L = (N_{\nu_L} - N_{\bar{\nu}_L})/N_\gamma$ have been obtained:

$$-0.071 < \eta_L < 0.054. \tag{2}$$

The fact that non diagonal neutrino magnetic moment μ_ν could induce a neutrino-antineutrino transition due to a helicity flip produced by the interaction of μ_ν with an external magnetic field is known since a long time. It has been called Spin-Flavor precession (SFP) effect. This effect was originally used to explain the solar neutrinos deficit [13, 14, 15, 24]. However, after the confirmation of mixing mass explanation by KamLAND [25], the SFP is used as a mechanism to constraint μ_ν [26].

The effect of a primordial magnetic field on baryogenesis have already been studied [44, 45, 46, 44] but it has been done using the standard model anomaly terms which violates $B + L$ quantum numbers and not through SFP process.

In this talk, we consider the implications of the neutrino SFP on $\Delta L = 2$ processes and leptogenesis induced by strong primordial magnetic fields [12]. In section II we briefly review the neutrino spin flavor precession, induced by the primordial magnetic fields. In section III the time dependent magnetic fields at early universe is discussed. The values of neutrino magnetic moments are discussed in section IV. Section V is devoted to study the effect of SFP on a possible lepton asymmetry previously generated . In section VI the effect of SFP on leptogenesis is studied. Finally our conclusions and remarks are given in section VII.

2. Neutrino Spin Flavor Precession

The assumption that neutrino magnetic moment could be an explanation to the deficiency of the solar neutrino flux through Spin Precession effect were exposed by Cisneros more than 40 years ago [13] and generalized later to the case of Majorana neutrinos [47]. It is well known that left-handed fermion with magnetic moment could be affected by the Spin Precession effect (SP) which induces in presence of magnetic field a transition from left to right handed fermions or inversely[14, 15, 16, 17, 18, 19, 20, 21, 22, 23]. For the Majorana neutrinos the diagonal components of magnetic moments vanish and the off-diagonal components are related by $-\mu_{e\mu} = \mu_{\mu e} \equiv \mu_\nu$ leading to processes violating flavors and lepton number. In order to find the probability of the $\nu_{eL} \to \nu^c_{\mu L}$ transition, we need to study the evolution of the chiral components of two flavors of neutrinos, which is described by a Schrödinger type equation [16].Since we are interested at early epochs of the universe ($T \sim 10^{11}$ GeV), we ignore the neutrino masses and the electron-neutron energy densities. Hence, the solution to the Schrödinger type equation [16] with an arbitrary magnetic field is given by

$$P(\nu_{eL} \to \nu^c_{\mu L}; t) = \sin^2 \left(\int_{t_0}^{t} \mu_\nu B_\perp(t')\, dt' \right) . \tag{3}$$

where $B_\perp(t)$ is the transverse magnetic field strength, t being the time appearing in the Schrödinger equation. This formula is valid for an arbitrary magnetic field profile $B_\perp(t)$. It is important to notice that the SFP will not stop at Electroweak breaking scale but will continue up to our days.

3. Time-dependent magnetic fields at Early Universe

The main constraints on SFP processes are coming from limits on primordial magnetic field at photon decoupling time obtained through observing microwave background radiation [48] which puts a limit on present time magnetic field to be smaller than $3 \times 10^{-9}\ G$ [27]. This limit should be translated into the primordial time assuming that the magnetic field evolution is given by

$$B(t) \simeq B(t_i) \left(\frac{a(t_i)}{a(t)} \right)^2 , \tag{4}$$

where $a(t)$ is the scale factor, assuming Friedman Robertson Walker dynamics for the Universe (for detailed discussion see ref. [29]). This means that the CMB limit on present value of the primordial magnetic field could be roughly translated into a limit of order $10^9 G$ for the primordial magnetic field at BBN time [28, 29], which correspond to a time of around 100 s after Big Bang Thus, it is crucial to translate this bound on the primordial magnetic fields at electroweak symmetry breaking scale and up to the scale of right-handed majorana neutrino decoupling (M_1 around $10^{11} GeV$), where the leptogenesis process takes place. At these times (which correspond to radiation domination era) , the scale factor is given by

$$a(t) \propto t^{1/2}. \tag{5}$$

Thus, the bound on magnetic field at electroweak scale is of order 10^{17} G and at M_1 scale is around $10^{27} G$. In this respect, we assume that our time-dependent magnetic field between the time associated to the scale of the heavy right-handed Majorana Neutrinos typically given by $M_1 \simeq 10^{11} GeV$ and the time (t_{EPT}), which corresponds to the time when the Electoweak Phase Transition (EPT) occurs, is given by:

$$B(t) \simeq B(t_{EPT}) \frac{t_{EPT}}{t} \tag{6}$$

where $B(t_{EPT}) \simeq 10^{17} G$.

4. Majorana neutrinos and magnetic moment

Clearly from eq. (3) and with primordial magnetic fields as described above, to have a significant Spin-Flavour precession effect it is necessary to have magnetic moments for the Majorana neutrinos not too small. Within the Standard Model, the left-handed majorana neutrinos get a magnetic moment through radiative corrections [30, 31, 32, 33, 34]:

$$\mu_\nu \approx 3.2 \times 10^{-19} \left(\frac{m_\nu}{1eV} \right) \mu_B \tag{7}$$

where μ_B is the Bohr magneton. The strongest constraints on μ_{ν_e} come from Reactor experiment as TEXONO [35] or GEMMA experiments [36]:

$$\mu_{\nu_e} \leq 7.4 \times 10^{-11} \mu_B (TEXONO) \tag{8}$$

$$\mu_{\nu_e} \leq 3.2 \times 10^{-11} \mu_B (GEMMA) \tag{9}$$

From solar neutrinos, it is possible to constraint the neutrino magnetic moments. Recently, BOREXINO [38] obtains the following bound:

$$\mu_{\nu_e} \leq 5.8 \times 10^{-11} \mu_B \tag{10}$$

For Majorana neutrinos, global fits on neutrino magnetic moments have been done using solar and reactor data and they obtained [37]:

$$\mu_{ij} \leq 2.0 \times 10^{-10} \mu_B \tag{11}$$

The experimental limits are far from the expected values for the neutrino magnetic moments as computed within Standard Model. The main difficulty to build models beyond Standard Model with large neutrino magnetic moment is due to the fact that a large magnetic moment is usually related to strong New Physics contribution to neutrino masses but this problems can be solved in many New Physics models where $\mu_{\nu_{ij}}$ as large as $10^{-12} \mu_B$ can be reached even for m_ν smaller than 1 eV [39, 40, 41, 42, 43]. In models beyond the standard model like MSSM, the magnetic moments of neutrinos are enhanced by several order of magnitudes, for example $\mu_{\nu_e} \sim 10^{-15} 10^{-16} \mu_B$, and $\mu_{\nu_\mu} \sim 10^{-12} 10^{-13} \mu_B$, while $\mu_{\nu_\tau} \sim 10^{-12} \mu_B$. Therefore, the neutrino magnetic moments can be a probe for the new physics beyond the standard model. For our numerical estimate of the SFP effects, we shall use the value of $10^{-12} \mu_B$ for the majorana neutrino magnetic moments.

5. Spin Flavor Precession and Lepton asymmetry

We now study the effect of spin-precession process on light, mainly left-handed, majorana neutrino assuming the existence of a time-dependent primordial magnetic fields given in Eq.(6). In order to include in the Boltzman equation the terms corresponding to the spin-precession effects, it is important to recall in two flavor case that the variation $\Delta N_{\nu_{1,2}}$ in $\nu_{1,2}$ number density due to SFP is given by

$$\Delta N_{\nu_1} = P(\nu_2^c \to \nu_1)N_{\nu_2^c} - P(\nu_1 \to \nu_2^c)N_{\nu_1} \tag{12}$$
$$\Delta N_{\nu_2} = P(\nu_1 \to \nu_2^c)N_{\nu_1} - P(\nu_2^c \to \nu_1)N_{\nu_2^c} \tag{13}$$

where the first term in Eq. (12) account for the number of ν_2^c's which have been changed into ν_1 and the second term is equal to the number of ν_1's which have been changed into ν_2^c. Similar equations can be built for $\Delta N_{\nu_{1,2}^c}$. Defining the lepton number density as

$$N_L \equiv N_{\nu_1} + N_{\nu_2} - N_{\nu_1^c} - N_{\nu_2^c} \ ,$$

and assuming CP is conserved ($i.e.$, $P(\nu_1 \to \nu_2^c) = P(\nu_1^c \to \nu_2)$), one gets

$$\Delta N_L = -2PN_L \tag{14}$$

where P is the probability of SFP given by Eq. (3). This approach can be easily extend to n_f flavors and we obtain

$$\frac{dN_L}{dt} = -2(n_f - 1)\frac{d}{dt}(PN_L) \, . \tag{15}$$

This equation represents a new contribution for the lepton asymmetry that will affect the leptogenesis scenario. Thus, the lepton number density $N(t)$ is given by

$$N_L(t) = \frac{N_L^0}{1 + 2(n_f - 1)P(\bar{\nu} \to \nu)}, \tag{16}$$

where N_L^0 is the initial lepton number density. It is clear that for probability $P(\bar{\nu} \to \nu) \simeq \mathcal{O}(1)$, the lepton asymmetry can be reducing respect to its initial value by a factor 1/5.

It is worth mentioning that for magnetic fields below 10^{14} G and due to limit on the neutrino magnetic moment [26] to be $\mu < 10^{-12}\mu_B$, one can easily check from Eq.(3) that the SFP process is irrelevant between the scale of right-handed neutrino decays (t_{M_1}) and electroweak symmetry breaking time (t_{EPT}). So for primordial magnetic field smaller than $10^{14}G$, SFP process will not affect directly the usual leptogenesis scenario. However as it will be shown explicitly below, it will continue to affect the lepton asymmetry even after the electroweak symmetry breaking, transforming it as a time-oscillating function. This is important as in usual leptogenesis scenario, the lepton and baryon asymmetry of the Universe are related through a simple relation which only depends on matter contents of the model. Even with a relatively weak primordial magnetic field (below 10^{14} G), this relation between η_L and η_B is lost.

6. Leptogenesis and Spin Flavor Precession

Here, we assume a strong primordial time-dependent magnetic field, given by Eq.(6), before electroweak phase transition and compatible with present limits on cosmological magnetic fields. In order to get the SFP effects on Leptogenesis standard scenario[1], we solve the Boltzman equation of the heavy right-handed (RH) majorana neutrino, N_1, that decays violating CP and

[1] for a detail description of Leptogenesis standard scenario see for instance ref. [7]

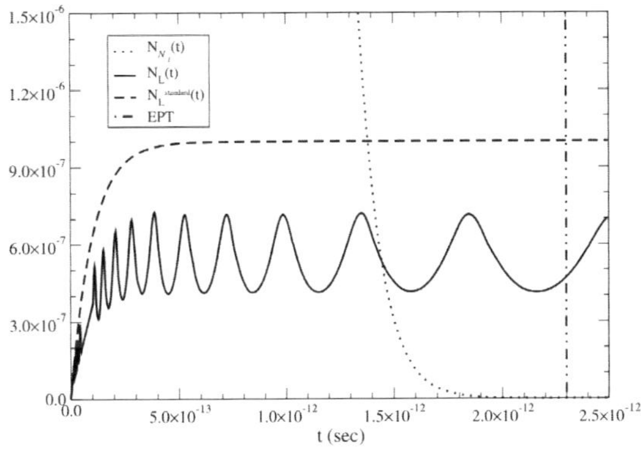

Figure 1. The continue oscillating line corresponds to η_L including SFP effects. The dash line is η_L as expected from standard leptogenesis scenario.The vertical dotted-dash line represents an approximate value for Electoweak Phase Transition Time (t_{EPT}). For $t > t_{EPT}$, the η_L is not anymore converted into η_B through $B + L$ violating sphalerons. The dotted line show the evolution of the heavy RH majorana neutrino density N_1

producing a lepton asymmetry through usual leptogenesis scenario. This lepton asymmetry is transformed into the Baryon Asymmetry of the Universe through anomalous $B + L$ violating sphaleron processes which are in equilibrium between $10^{12}GeV > T > 100\ GeV$. For simplicity and in order to clearly see the SFP effects on Boltzman equations, we assume that the $\Delta L = 1$ scattering processes in Boltzman equation for heavy RH majorana neutrinos are out of equilibrium and that the only relevant terms is the one describing the heavy RH neutrino decays and inverse decays. Also, for the N_L Boltzman equation, we assume that all $\Delta L \neq 0$ processes induced by heavy neutrinos are out of equilibrium and are not able to wash out any produced lepton asymmetry. Within these approximation, the basic equations for leptogenesis including SFP effects are given by

$$\begin{aligned} \frac{dN_{N_1}}{dt} &= -\Gamma_D N_1 \\ \frac{dN_L}{dt} &= \epsilon\Gamma_D N_{N_1} - 2(n_F - 1)\frac{d(PN_L)}{dt} \end{aligned} \tag{17}$$

where Γ_D represent the Direct and Inverse Decay and N_1 is the heavy RH Majorana neutrino density. From [7], we use $\epsilon = 10^{-6}$ and Γ_D is given by

$$\Gamma_D = \frac{1}{8\pi}\frac{m_1 M_1}{v^2}M_1\frac{K_1(z)}{K_2(z)}. \tag{18}$$

where $K_i(z)$ are the Bessel functions, and m_1 is the effective light neutrino mass, v is the usual electroweak symmetry breaking scale and M_1 is the heavy RH neutrino mass[7]. The results of integrating these equations are shown in Fig 1.

7. Conclusion

We have studied the impact of the neutrino spin flavor precession, induced by coherent primordial magnetic fields, on the lepton asymmetry and leptogenesis process. In order to implement the SFP leptogenesis we assumed a strong primordial magnetic fields and sizeable neutrino magnetic moment in addition to the usual assumptions of standard leptogenesis, required to generate the Baryon Asymmetry of the Universe in particular the wash out processes are out of equilibrium.

We have shown that contrary to what could be naively expected from the weakness of the extra and intra galactic magnetic fields at present time, primordial magnetic fields in Early Universe could be large enough to significantly affect Leptogenesis scenario. With such strong magnetic field at electroweak symmetry breaking time, we have shown that the SFP effects reduce the lepton asymmetry by around a factor 50% and increase the uncertainties on the produced BAU as the uncertainties on the electroweak phase transition time which corresponds to the freezing of the BAU are important. Even for magnetic fields too weak to modify Leptogenesis scenario, their presence induces an oscillating behavior for the lepton asymmetry at later stage in the History of the Universe, leading to lose the relation between Lepton and Baryon asymmetry as usually given in Leptogenesis models. A profile for the magnetic fields up to time around 100 seconds after Big Bang is needed in order to perform more precise numerical. It is also important to stress that the hypothesis of coherent primordial magnetic fields can be easily relaxed once the distribution of the primordial magnetic fields is known. Within this SFP leptogenesis scenario, an important consequence of a no-coherent distribution of primordial magnetic fields is to induce the formation of region of the Universe with different values for the lepton asymmetry.

Acknowledgements
The work of S. K. is partially supported by the ICTP grant AC-80. D.D. and R.C. are grateful to Conacyt (México), DAIP project (Guanajuato University) and PIFI (Secretaria de Educacion Publica, México) and organizers for financial support.

References

[1] G. Steigman. *Ann.Rev.Astron.Astrophys.*, 14:339–372, 1976.
[2] Gary Steigman. *JCAP*, 0810:001, 2008.
[3] M. Fukugita and T. Yanagida. *Phys.Lett.*, B174:45, 1986.
[4] M.A. Luty. *Phys.Rev.*, D45:455–465, 1992.
[5] Esteban Roulet, Laura Covi, and Francesco Vissani. *Phys.Lett.*, B424:101–105, 1998.
[6] W. Buchmuller and M. Plumacher. *Phys.Lett.*, B431:354–362, 1998.
[7] W. Buchmuller, P. Di Bari, and M. Plumacher. *Annals Phys.*, 315:305–351, 2005.
[8] V.A. Kuzmin, V.A. Rubakov, and M.E. Shaposhnikov. *Phys.Lett.*, B155:36, 1985.
[9] V.A. Matveev, V.A. Rubakov, A.N. Tavkhelidze, and M.E. Shaposhnikov. *Sov.Phys.Usp.*, 31:916–939, 1988.
[10] Gary Steigman. arXiv:1008.4765 [astro-ph.CO].
[11] E. Castorina, U. Franca, M. Lattanzi, J. Lesgourgues, G. Mangano, A. Melchiorri and S. Pastor, arXiv:1204.2510 [astro-ph.CO].
[12] J. Barranco, R. Cota, D. Delepine and S. Khalil, arXiv:1205.1250 [hep-ph].
[13] Arturo Cisneros. *Astrophys.Space Sci.*, 10:87–92, 1971.
[14] L.B. Okun, M.B. Voloshin, and M.I. Vysotsky. *Sov.J.Nucl.Phys.*, 44:440, 1986.
[15] L.B. Okun, M.B. Voloshin, and M.I. Vysotsky. *Sov.Phys.JETP*, 64:446–452, 1986.
[16] C. -S. Lim and W. J. Marciano, Phys. Rev. D **37**, 1368 (1988).
[17] R. Barbieri and G. Fiorentini, Nucl. Phys. B **304**, 909 (1988).
[18] E. K. Akhmedov, Sov. J. Nucl. Phys. **48**, 382 (1988) [Yad. Fiz. **48**, 599 (1988)].
[19] E. K. Akhmedov, Phys. Lett. B **213**, 64 (1988).
[20] A. Y. .Smirnov, Phys. Lett. B **260**, 161 (1991).
[21] C. Aneziris and J. Schechter, Int. J. Mod. Phys. A **6**, 2375 (1991).
[22] H. Nunokawa and H. Minakata, Phys. Lett. B **314**, 371 (1993).
[23] E. K. Akhmedov, hep-ph/9705451.
[24] O.G. Miranda, Carlos Pena-Garay, T.I. Rashba, V.B. Semikoz, and J.W.F. Valle. *Nucl.Phys.*, B595:360–380, 2001.
[25] J. Barranco, O.G. Miranda, T.I. Rashba, V.B. Semikoz, and J.W.F. Valle. *Phys.Rev.*, D66:093009, 2002.
[26] O.G. Miranda, T.I. Rashba, A.I. Rez, and J.W.F. Valle. *Phys.Rev.Lett.*, 93:051304, 2004.
[27] Dario Grasso and Hector R. Rubinstein. *Phys.Rept.*, 348:163–266, 2001.
[28] J.J. Matese and R.F. O'Connell. *Phys.Rev.*, 180:1289–1292, 1969.
[29] Alejandra Kandus, Kerstin E. Kunze, and Christos G. Tsagas. *Phys.Rept.*, 505:1–58, 2011.
[30] W. J. Marciano and A. I. Sanda, Phys. Lett. B **67**, 303 (1977).
[31] B. W. Lee and R. E. Shrock, Phys. Rev. D **16**, 1444 (1977).

Journal of Physics: Conference Series **468** (2013) 012003 doi:10.1088/1742-6596/468/1/012003

[32] K. Fujikawa and R. Shrock, Phys. Rev. Lett. **45**, 963 (1980).
[33] P. B. Pal and L. Wolfenstein, Phys. Rev. D **25**, 766 (1982).
[34] R. E. Shrock, Nucl. Phys. B **206**, 359 (1982).
[35] H. T. Wong *et al.* [TEXONO Collaboration], Phys. Rev. D **75**, 012001 (2007) [hep-ex/0605006].
[36] A. G. Beda, E. V. Demidova, A. S. Starostin, V. B. Brudanin, V. G. Egorov, D. V. Medvedev, M. V. Shirchenko and T. .Vylov, Phys. Part. Nucl. Lett. **7**, 406 (2010) [arXiv:0906.1926 [hep-ex]].
[37] W. Grimus, M. Maltoni, T. Schwetz, M. A. Tortola and J. W. F. Valle, Nucl. Phys. B **648**, 376 (2003) [hep-ph/0208132].
[38] C. Arpesella *et al.* [Borexino Collaboration], Phys. Rev. Lett. **101**, 091302 (2008) [arXiv:0805.3843 [astro-ph]].
[39] M. B. Voloshin, Sov. J. Nucl. Phys. **48**, 512 (1988) [Yad. Fiz. **48**, 804 (1988)].
[40] S. M. Barr, E. M. Freire and A. Zee, Phys. Rev. Lett. **65**, 2626 (1990).
[41] K. S. Babu and R. N. Mohapatra, Phys. Rev. Lett. **63**, 228 (1989).
[42] K. S. Babu and R. N. Mohapatra, Phys. Rev. Lett. **64**, 1705 (1990).
[43] M. Leurer and N. Marcus, Phys. Lett. B **237**, 81 (1990).
[44] Massimo Giovannini and M.E. Shaposhnikov. *Phys.Rev.*, D57:2186–2206, 1998.
[45] V.B. Semikoz and J.W.F. Valle. *JHEP*, 0803:067, 2008.
[46] V.B. Semikoz, D.D. Sokoloff, and J.W.F. Valle. *Phys.Rev.*, D80:083510, 2009.
[47] J. Schechter and J.W.F. Valle. *Phys.Rev.*, D24:1883–1889, 1981.
[48] Kiyotomo Ichiki, Keitaro Takahashi, and Naoshi Sugiyama. *Phys.Rev.*, D85:043009, 2012.

Neutrino-Electron Scattering: Charge Radius and Effective Couplings

A. Gutiérrez-Rodríguez, A. González-Sánchez and M. C. Blanco
Facultad de Física, Universidad Autónoma de Zacatecas
Apartado Postal C-580, 98060 Zacatecas, México.

M. A. Hernández-Ruíz
Unidad Académica de Ciencias Químicas, Universidad Autónoma de Zacatecas
Apartado Postal C-585, 98060 Zacatecas, México.

E-mail: `alexgu@fisica.uaz.edu.mx`

Abstract. In this work the neutral-current scattering cross-section for neutrinos on electrons is calculated assuming that a massive Dirac neutrino is characterized by a phenomenological parameters, a charge radius $\langle r^2 \rangle$ and the right-handed currents are present in the framework of a Left-Right symmetric model (LR). Using the CHARM II result for the charge radius of the muon-neutrino $|\langle r^2 \rangle| < 6.0 \times 10^{-33}\ cm^2$, we place a bound on $-7.9 \times 10^{-33}\ cm^2 \leq \langle r^2 \rangle_{LR} \leq 7.9 \times 10^{-33}\ cm^2$. We discuss the relationship between the electron neutral couplings $g_V^{\nu e}$ and $g_A^{\nu e}$ and the LR model parameters.

1. Introduction

In general, a photon may couple to charged leptons through its electric charge, magnetic dipole moment (MM), electric dipole moment (EDM) and the anapole moment (AM). This coupling may be parametrized using a matrix element in which the usual γ^α is replaced by a more general Lorentz-invariant form [1]:

$$\Gamma^\mu = F_Q(q^2)\gamma^\mu + F_M(q^2)i\sigma^{\mu\nu}q_\nu + F_E(q^2)\sigma^{\mu\nu}q_\nu\gamma_5 + F_A(q^2)(q^2\gamma^\mu - q^\mu \not{q})\gamma_5, \quad (1)$$

where $F_{Q,M,E,A}(q^2)$ are the electromagnetic form factors of the neutrino, corresponding to the charge radius, MM, EDM and AM, respectively.

Electromagnetic properties of neutrinos are of fundamental importance and serve as a probe of physics beyond the SM. Several authors have shown that the charge radius of the neutrino is not a physical quantity [2], as demonstrated by the fact that it is gauge-dependent [3]. However, other authors claim that they can extract a gauge-independent neutrino charge radius which is, therefore, a physical observable. A definition of the neutrino charge radius that satisfies physical requirements, i.e. it is a physical observable, has recently been provided in the framework of the Pinch Technique formalism [4].

In this paper, we start from a Left-Right symmetric model (LR) [5] and assuming that a massive Dirac neutrino is characterized by a phenomenological parameter, a charge radius $\langle r_\nu^2 \rangle$ we calculate the cross-section of the process $\nu_\mu e^- \to \nu_\mu e^-$. We also estimate bounds on the charge radius of the muon-neutrino in the framework of the LR model.

2. The Muon-neutrino Electron Cross-Section

In this section we obtain the corresponding amplitude for the process

$$\nu_\mu(k_1) + e^-(p_1) \rightarrow \nu_\mu(k_2) + e^-(p_2), \tag{2}$$

mediated by the photon γ and the neutral gauge bosons Z_L and Z_R. We assume that a massive Dirac neutrino is characterized by a phenomenological parameters, a charge radius $\langle r_\nu^2 \rangle$. Therefore, the expression for the cross-section of the process $\nu_\mu e^- \rightarrow \nu_\mu e^-$ is give by

$$\begin{aligned} \sigma_T^{LR} &= \frac{G_F^2 m_e E_\nu}{2\pi}\{2\delta^2 + 2\delta(P+S) + \frac{(P+S)^2 + (Q+R)^2}{2} \\ &+ \frac{1}{3}[2\delta^2 + 2\delta(P-S) + \frac{(P-S)^2 + (Q-R)^2}{2}]\}, \end{aligned} \tag{3}$$

where P, Q, R, S are given by

$$\begin{aligned} P &= (A + 2B + C)g_V, \\ Q &= (-A + C)g_A, \\ R &= (-A + C)g_V, \\ S &= (A - 2B + C)g_A. \end{aligned} \tag{4}$$

The constants A, B and C depend only on the parameters of the LR model

$$\begin{aligned} A &= (c_\phi - \frac{s_W^2}{r_W}s_\phi)^2 + \gamma(\frac{s_W^2}{r_W}c_\phi + s_\phi)^2, \\ B &= (c_\phi - \frac{s_W^2}{r_W}s_\phi)(-\frac{c_W^2}{r_W}s_\phi) + \gamma(\frac{s_W^2}{r_W}c_\phi + s_\phi)(\frac{c_W^2}{r_W}c_\phi), \\ C &= (\frac{c_W^2}{r_W}s_\phi)^2 + \gamma(\frac{c_W^2}{r_W}c_\phi)^2, \\ \gamma &= (\frac{M_{Z_L}}{M_{Z_R}})^2, \\ \delta &= \frac{\sqrt{2}\pi\alpha}{3G_F}\langle r^2 \rangle, \end{aligned} \tag{5}$$

where $c_\phi = \cos\phi$, $s_\phi = \sin\phi$, $c_W = \cos\theta_W$, $s_W = \sin\theta_W$, $r_W = \sqrt{\cos 2\theta_W}$ and M_{Z_L}, M_{Z_R} are the masses of the light and heavy massive neutral vector bosons that participate in the interaction. γ together with ϕ are the two new parameters that are introduced in the LR model, while $\langle r^2 \rangle$ is the charge radius of the muon-neutrino. Bounds on this quantity are reported in the literature [6]. We can in addition take the limit $\phi = 0$ and $M_{Z_2} \rightarrow \infty$ and get the standard model cross-section.

From the cross-section expression (3), we obtain the interference cross-section which is given by

$$\sigma_I^{LR} = \frac{2G_F^2 m_e E_\nu \delta}{3\pi}(2P + S). \tag{6}$$

We rewrite the interference cross-section in order to identify the neutrino charge radius in the LR model

$$\langle r^2\rangle_{LR} = \langle r^2\rangle[1+(\frac{2g_V a_W + g_A b_W}{2g_V + g_A})\gamma + (\gamma - 1)(\frac{2g_V + g_A b'_W}{2g_V + g_A})s_\phi c_\phi], \quad (7)$$

where $\langle r^2\rangle_{LR}$ is the charge radius of the muon-neutrino in the LR model and $\langle r^2\rangle$ is the charge radius in the minimal extension of the standard model.

The total cross-section Eq. (3) is written in such a way that we can express the theoretical predictions for the electron couplings constant $(g_V^{\nu e})_{LR}$ and $(g_A^{\nu e})_{LR}$ such that give the SM couplings in the limit $\phi = 0$ and $M_{Z_2} \to \infty$:

$$(g_A^{\nu e})_{LR}[M_{Z_L}, M_{Z_R}, \phi, \delta, \sin\theta_W, g_A^{\nu e}] = [\delta f_3 + \frac{1}{2}(f_1 f_3 + f_2^2)](g_A^{\nu e})_{SM}, \quad (8)$$

$$(g_V^{\nu e})_{LR}[M_{Z_L}, M_{Z_R}, \phi, \delta, \sin\theta_W, g_V^{\nu e}] = \delta f_3 + \frac{1}{2}(f_1 f_3 + f_2^2)(g_V^{\nu e})_{SM}. \quad (9)$$

In these expressions, with the limits $\delta = 0$, $\phi = 0$ and $M_{Z_R} \to \infty$, the SM couplings are recovered.

3. Results and Conclusions

In this section, we present the numerical results obtained for the charge radius of the muon-neutrino in the framework of a Left-Right symmetric model $\langle r^2\rangle_{LR}$, and the electron couplings constants $(g_V^{\nu e})_{LR}$ and $(g_A^{\nu e})_{LR}$. For the mixing angle ϕ between Z_L and Z_R, we use the reported data in Ref. [7]: $-1.66 \times 10^{-3} \leq \phi \leq 1.22 \times 10^{-3}$, with a 90% C.L.

In order to estimate a limit on the charge radius of the muon-neutrino $\langle r^2\rangle_{LR}$ in the framework of the Left-Right symmetric model, we plot the expression (7) in Fig. 1. In this figure, we show the allowed region for $\langle r^2\rangle_{LR}$ as a function of ϕ with 90% C.L. The allowed region is the rectangle band that is a result of both factors in Eq. (7). In this figure, the second factor gives the rectangle band while $\langle r^2\rangle$ gives the band width. This analysis was done using the experimental value for $\langle r^2\rangle$ reported by CHARM II [6] with a 90% C.L. In the same figure, we show the $\langle r^2\rangle$ $(\phi = 0)$ result at 90% C.L. with the dashed horizontal lines. The allowed region in the LR model (dotted line) for $\langle r^2\rangle_{LR}$ is wider than the one for the $\langle r^2\rangle$, and is given by

$$-7.9 \times 10^{-33}\ cm^2 \leq \langle r^2\rangle_{LR} \leq 7.9 \times 10^{-33}\ cm^2, \quad 90\%\ \text{C.L.}, \quad (10)$$

whose value is quite close to that reported by other authors [6].

Fig. 2 shows the charge radius $\langle r^2\rangle_{LR}$ as a function of the LR parameters ϕ and M_{Z_R}. This figure shows a strong dependence of the charge radius with respect to the model parameters.

In Fig. 3 we have plotted $(g_A^{\nu e})_{LR}$ from Eq. (9) as a function of the LR parameters ϕ and M_{Z_R}. We have chosen the range $-1.66 \times 10^{-3} \leq \phi \leq 1.22 \times 10^{-3}$ and $300 \leq M_{Z_R} \leq 800\ GeV$ where ϕ is measured in radians.

In Fig. 4 we have plotted $(g_V^{\nu e})_{LR}$ from Eq. (8) as a function of ϕ and M_{Z_R}. The range of variation for the LR parameters is the same as in Fig. 3. In this case the experimental value, $(g_V^{exp})_{LR} = -0.035$ is reached for small values of M_{Z_R} and ϕ. The effect of ϕ and M_{Z_R} on $(g_A^{\nu e})_{LR}$ and $(g_V^{\nu e})_{LR}$ are similar.

In summary, we have estimated bounds that can be derived from the muon-neutrino electron scattering. Our bounds on the neutrino charge radius $\langle r^2\rangle_{LR}$ and the electron couplings constants $(g_V^{\nu e})_{LR}$ and $(g_A^{\nu e})_{LR}$ are consistent with those reported in the literature and in some cases, improve the existing bounds. However, new experiments dedicated to the detailed study of electron (anti) neutrino interactions with matter, for example the reactor MUNU, as well as radioactive sources of neutrinos such as the BOREXINO detector, should be able to improve existing limits on the neutrino charge radius, magnetic moment and other parameters.

Acknowledgments

We acknowledge support from CONACyT, SNI and PROMEP (México).

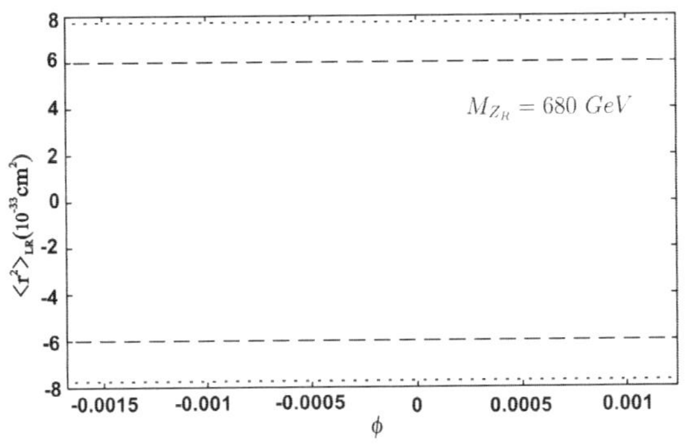

Figure 1. Allowed region for $\langle r^2 \rangle_{LR}$ as a function of the mixing angle ϕ with the value $\langle r^2 \rangle$.

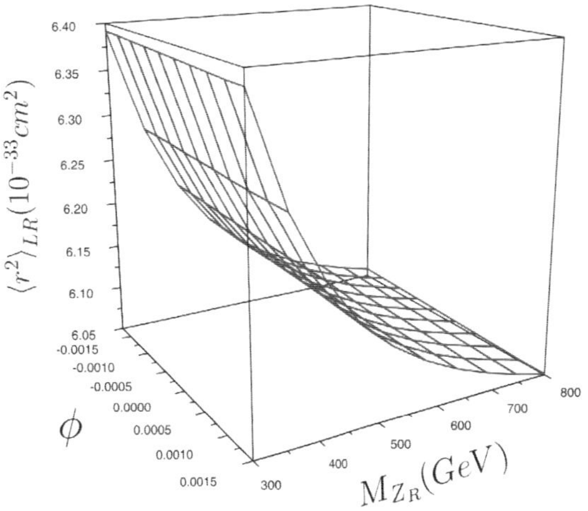

Figure 2. Plot of $\langle r^2 \rangle_{LR}$ as a function of the LR parameters ϕ and M_{Z_R}.

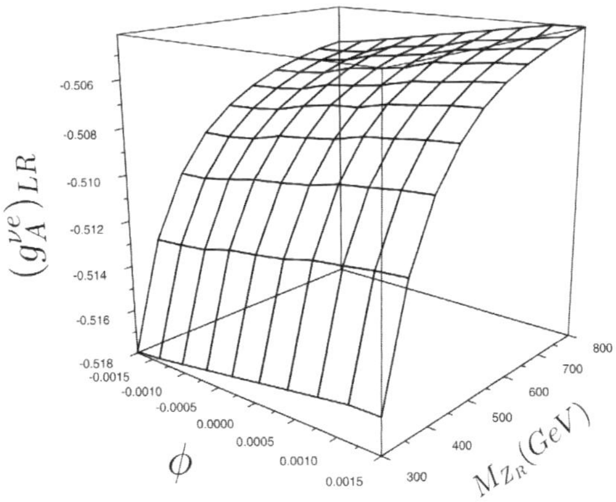

Figure 3. Plot of $(g_A^{\nu e})_{LR}$ as a function of the LR parameters ϕ and M_{Z_R}.

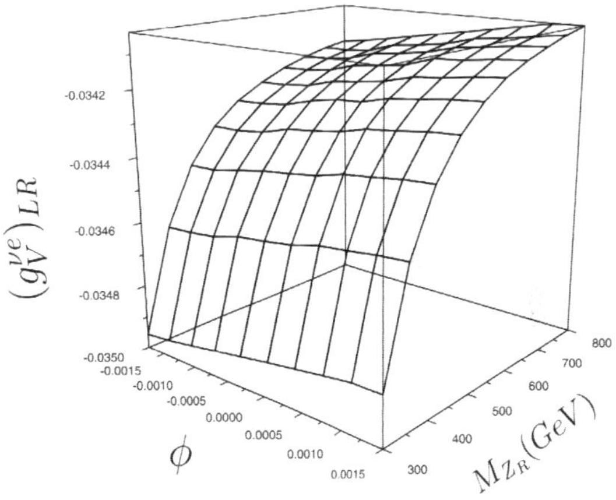

Figure 4. Plot of $(g_A^{\nu e})_{LR}$ as a function of the LR parameters ϕ and M_{Z_R}.

References

[1] B. Kayser, **Phys. Rev. D 26**, (1982) 1662.
[2] W. A. Bardeen, R. Gastmans and B. E. Lautrup, Nucl. Phys. B **46**, 1972, 319.
[3] B. W. Lee and R. E. Shrock, Phys. Rev. D **16**, 1977, 1444.
[4] J. M. Cornwall and J. Papavassiliou, Phys. Rev. D **40**, 1989, 3474.
[5] J. C. Pati and A. Salam, Phys. Rev. D **10**, 1974, 275.
[6] P. Vilain *et al.* [CHARM-II Collaboration], Phys. Lett. B **345**, 1995, 115.
[7] A. Gutierrez-Rodriguez, M. A. Hernandez-Ruiz and M. A. Perez, Int. J. Mod. Phys. A **22**, 2007, 3493; A. Gutierrez-Rodriguez, M. A. Hernandez-Ruiz, M. A. Perez and F. Perez-Vargas, hep-ph/0702076; A. Gutierrez-Rodriguez, M. A. Hernandez-Ruiz, M. A. Perez and F. Perez-Vargas, Mod. Phys. Lett. A **24**, 2009, 135.

Journal of Physics: Conference Series **468** (2013) 012005 doi:10.1088/1742-6596/468/1/012005

Lorentz Invariance Violation on UHECR propagation

H Martínez, A Pérez-Lorenzana

Centro de Investigación y de Estudios Avanzados del IPN, México, DF, México.

E-mail: hmartinez@fis.cinvestav.mx, aplorenz@fis.cinvestav.mx

Abstract. In the present work possible consequences of the Lorentz symmetry breaking are presented. Those consequences are encoded through the modification of dispersion relations which are applied to the processes of pair and pion photo production generated at UHECR scale. The produced physics from such modifications contains scenarios where the expected photo production thresholds under Lorentz symmetry are shifted; the other scenarios predict that the cosmic ray flux will be recover at higher energies due to the suppression of the production processes. This phenomenon is repeated for different components of the diffuse photon background and generates opacity bands in distinct parts of the energy spectrum; we expose scenarios that have not been considered previously by other authors. Moreover, in contrast with previous studies, in this work the scenarios generated by the Lorentz symmetry breaking are measured by a generic parameter, unique for each process, independent of the correction origin (E^2 or p^2) in the dispersion relations, as well as the nature of the particle involved and possible corrections to the relations of conservation energy-momentum laws. Additionally, the Pierre Auger result for the pion photo production is used to estimate the energy scale parameter of the Lorentz symmetry breaking.

1. Introduction

Many Quantum Gravity theories suggest the breaking of Lorentz Invariance (LI) with the strength of the effects increasing with energy [1]. If LI is broken in form of non-standard dispersion relations for various particles, absorption and energy loss processes for cosmic radiation interactions can be modified. Conversely, experimental confirmation that such processes occur at the expected thresholds would allow to put strong constraints on such LI breaking effects.

In the next section we expose the general background that has been used in the forward analysis. In section 3, we introduce the method. In section 4, we present the effects of the LI violation (LIV) for the pair and pion photo production by n-order modification. In section 5, we present the special case $n = 1$ and also a generic example, then we explore an experimental restrictions for the LIV pion photo production. Finally in section 6, conclusions are presented.

2. Diffuse γ-background

Since the early universe, several phenomena had contributed for the generation of a diffuse gamma background, which fills the space in an almost homogeneous way. One of the most known components of this background, and probably the most important in cosmology, is the Cosmic Microwave Radiation (CMB). Other relevant radiations are the Far Infrared Background (FIRB) and Radio Background (RB). The mean energies for those photons are $\omega_{CMB} \approx 6 \times 10^{-4} eV$, $\omega_{FIRB} \approx 10^{-2} eV$ and $\omega_{RB} \approx 4 \times 10^{-9} eV$.

Journal of Physics: Conference Series **468** (2013) 012005 doi:10.1088/1742-6596/468/1/012005

The diffuse background has a certain probability of interacting with cosmic rays. One can use Special Relativity (SR) to compute the minimum energy needed by the colliding particles to reproduce the pair and pion photo production ($\gamma_{HE}\gamma_b \to e^+e^-$; $p\gamma_{CMB} \to \Delta \to p\pi^0$, $p\gamma_{CMB} \to \Delta \to n\pi^+$) and they are: $\omega_{th} = m_e^2/\omega_b$ and $E_{p_ith} \approx [m_\pi(2m_p + m_\pi)]/4E_\gamma$. Then, for the diffuse γ-background components stated before, the thresholds predictions by this SR approximations are $\omega_{th_{FIRB}} \sim 3 \times 10^{13}$ eV, $\omega_{th_{CMB}} \sim 5 \times 10^{14}$ eV, $\omega_{th_{Rb}} \sim 6 \times 10^{19}$ eV and $E_{th_{FIRB}} \sim 7 \times 10^{18}$ eV, $E_{th_{CMB}} \sim 1 \times 10^{20}$ eV $E_{th_{RB}} \sim 2 \times 10^{25}$ eV. The CMB pion photo production is very important for UHECR study since it predicts a cosmic ray flux suppression at $\sim 10^{20} eV$ (GZK cutoff). The Pierre Auger Observatory has already measured that there is a cosmic ray flux suppression at 4×10^{19} eV [2]. Auger has also set upper limits to the UHC-γs flux: 0.4 %, 0.5 %, 1%, 2.6 % and 8.9% above 1,2,3,4 and 10 EeV [3]. In an Invariant Lorentz regime, these processes are examples of low energy processes in the mass center which are boosted at the laboratory system with extremely high Lorentz factors.

3. Lorentz Invariance Violation (LIV)

LI is the invariance under the Lorentz group transformations. The symmetry under boost says that if a particle (or any system) has energy and momentum $P^\mu = (E, \vec{p})$ in a system C, then the energy and momentum $P'^\mu = (E', \vec{p'})$ for the same particle in the system C' will be given by Lorentz transformations, where $S := P_\mu P^\mu = E^2 - p^2 = m^2$ is a Lorentz invariant and define the well-known dispersion relation.

A common way to break the Lorentz symmetry is adding no LI terms in the Lagrangian in order to modify the dispersion relation. Since it can be done by modifying both the energy or the momenta, without loss of generality we can expand the new dispersion relation as

$$E^2 - p^2 - m^2 = A^2 \left(\epsilon(0) + \epsilon'(0) \left(\frac{A}{M}\right) + \epsilon''(0) \left(\frac{A}{M}\right)^2 + \dots \right), \tag{1}$$

where ϵ parametrizes the LIV in such way that it returns the LI regime at low energies and it is proposed as a function of (A/M), where A can take the form of E or p and M parametrizes the LIV energy scale. As a conclusion, a new general class of dispersion relation can be proposed

$$E^2 - p^2 = m^2 + \alpha_n A^{n+2}, \tag{2}$$

where $\alpha_n := \epsilon^{(n)}/M^n \simeq 1/M^n$. Equation (2) is a general form for many cases, that one can find on literature, that break LI by a dispersion relation modification (see for instance [1],[4],[5]).

4. LIV processes

In general, LI is expected to be broken at very high energy and it can be done by modifying the dispersion relation. Additionally, in most cases a special frame is needed (CMB-frame), since there is no more Lorentz symmetry, however, it is possible to modify the boost action in order to save the invariance under inertial frames [6]. These corrections have consequences to the energy-momentum conservation laws. In order to include such modifications on the analysis, we propose that the energy-momentum laws are corrected by a lineal LIV coefficients that could be functional dependent on $\{E, p\}$. Then, we use the inelasticity, K, in two particles collision where $E_{1_{final}} := K(E_{1_{initial}} + \omega_b + \Xi_E)$ and $E_{2_{final}} := (1 - K)(E_{1_{initial}} + \omega_b + \Xi)$, where Ξ_E is a LIV term and ω_b the energy photon from the diffuse background. We use the system where the collision is head on; since the UHECR has higher energy than the photon background, the final momentums are collinear. Additionally, we introduce the ultra relativistic regime approximation $m \ll \{E, p\} \ll M$ and $\omega_b \ll \{E_{proton}, \omega\}$. We also demand only first order LIV coefficients.

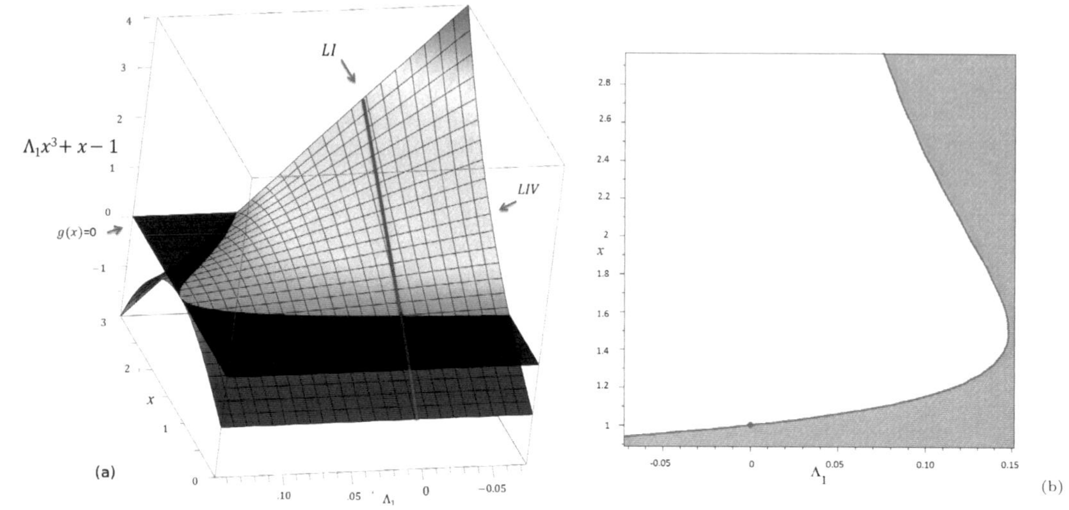

Figure 1. LIV scenarios. On the left the behavior of $g(x) = \Lambda_1 x^3 - x + 1$ for several Λ. At right, the black plane that points the zeros of $g(x)$ for a given range of Λ's.

Henceforth, for general modifications of n-order in the dispersion relation (2), the energy threshold equation has the form

$$\Lambda_n x^{n+2} - x + 1 = 0 \tag{3}$$

where $x = E_{th}^{VIL}/E_{th}^{IL}$ and $\Lambda_n := B(E_{th}^{IL})$; λ_n, is the energy independent LIV parameter and both B and λ_n are different for each process. As we expected, if $\lambda_n = 0$, LI regime is recover. In both cases, equation (3) is independent from particle nature, the functional form of A in equation (2) and Ξ, the last one means that the results do not depend on lineal corrections of the energy-momentum laws. For the pair production case we denote $\Lambda_n = \Omega_n(\alpha_{\gamma_{UHE}}, \alpha_\gamma, \alpha_{e^\pm}) := \omega_0^{n+1}/(4\omega_b)\beta_n$ and $\Lambda_n = \Upsilon_n(\alpha_p, \alpha_\gamma, \alpha_\pi)$ for pion photo production.

5. The $n = 1$ case

Each value of Λ_1 produces a different scenario, which describe, the behavior of the photo production processes. Figure (1a) displays in color the function $g(x) = \Lambda_1 x^3 - x + 1$ for $\Lambda_1 \neq 0$, these are all the possible LIV scenarios for $n = 1$. The red line is the expected physics on LI regime ($\Lambda_1 = 0$). The black plane is the intersection plane with $g(x) = 0$, and is also exposed in Figure (1b). The red line cut the black plane at $x = 1$, so $E_{th}^{VIL} = E_{th}^{IL}$. For the LIV scenarios one can distinguish two families by the sign of Λ_1. If $\Lambda_1 < 0$, each $g(x)$ scenario produces a threshold by the only intersection with the plane $g(x) = 0$, like the LI case. If $\Lambda_1 > 0$, $\Lambda_1 < 4/27$ by reality condition, $g(x)$ intersects two times the plane $g(x) = 0$. These solutions represent: the first (at lower energy) the energy threshold for the processes and the second (at higher energy) the value at which the processes will be forbidden again, this phenomena produces *opacity bands*. That means, that in these bands UHECR protons and photons are free of the photo production processes produced by the interactions with the DB. In this sense they can propagate freely in the universe. Additionally, since we demand real values of Λ_1, $\Lambda_1 < 4/27$ so $x \in [1, 1.5)$ and then, there is an upper limit for all the lower energy thresholds of the opacity bands.

In Figure (2) the case $\beta_1 = 1.8 \times 10^{-48} eV^{-1}$ for pair production is exposed. Since $(4/27)(4\omega_b/\omega_0^2) > \beta > 0$ the energy threshold equations has two solutions, so LIV produces *opacity bands*. Each one points out the energy region allowed for the interaction of the UHE-γs

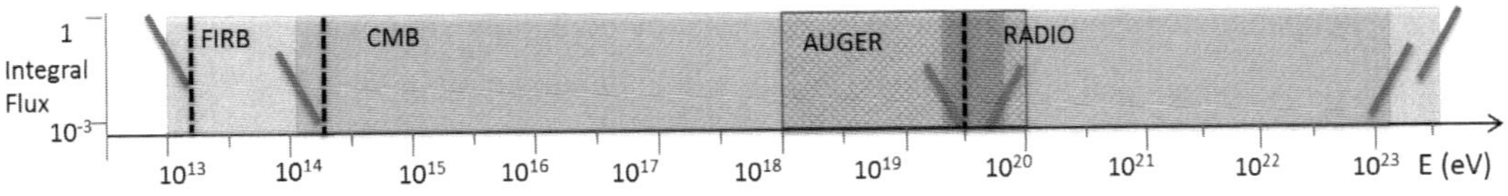

Figure 2. The figure shows the case for $\beta_1 = 1.8 \times 10^{-48} eV^{-1}$ in order to expose the opacity bands for each diffuse gamma background component. The bands are distributed over a wide energy range. The measure of one of these thresholds is enough to predict the location of the others.

with the different components of the DB. The doted lines are the upper limit that can reach the lowers walls of the bands. Outside each band, since the production process is forbidden for each background component, *the universe is transparent to the* γs *propagation.* The red lines are the slope of the expected flux of the incoming photons in the opacity band. Since pair production is an absorption process, the lines are expected to be pronounced. Dotted region indicates the energy range where Auger has set upper limits for the photon flux.

An extreme scenario would take place if $\beta_1 = 1.1 \times 10^{-32}\ eV^{-1}$, the RB threshold for photo pair production disappear since the solutions to the RB threshold equation has complex solutions. For the cases where $\Omega_1 = 4(\omega_b/\omega_0^2)\beta_1 \leq 0$ the opacity region grows indefinitely.

If Auger reported energy flux suppression, at 4×10^{19} eV, is taken as the $E_{p_{ith_{CMB}}}$, we can use the $n = 1$ results to compute the LIV parameter for CMB, which has the value $\Upsilon_{1_{CMB}} = -8.4$. Once Υ_1 is known, the energy thresholds for the other components for pion photo production are expected to be found at $E_{p_{ith_{FIRB}}} \approx 7 \times 10^{18} eV$ and $E_{p_{ith_{RB}}} \approx 5 \times 10^{19} eV$. In the very particular case where $\alpha_p = \alpha_\gamma = \alpha_\pi$, this means $|\alpha| \simeq 10^{-50} eV^{-1}$ and a LIV scale $M \geq 10^{50} eV$.

6. Conclusions

LIV models have generated observable astrophysical consequences which are measurable with the current state of the CR study. In this work it was found that under ultra relativistic regime, first order LIV terms and $\omega_b \ll E_{UHECR}$, the modifications given in equation (2), produce independent results from the particle nature, the functional form of A (E, p), and lineal corrections on the energy-momentum laws for the threshold analysis. We also found that the physics from LIV modifications contains scenarios where the expected photo productions thresholds can change drastically. These scenarios are measured by a generic parameter. In the scenarios where $E_{th}^{VIL} > E_{th}^{IL}$, it is predicted that the CR flux will be recovered at higher energies due to the presence of a second threshold which suppresses the production process. Then, opacity bands are generated. Hence, bands have direct consequences on the CR propagation through the universe. In second scenarios, the energy threshold is just moved to low energies. It was found that for both processes there is a limit value that each lower wall of the opacity bands could take. This limit exists for each one of the diffuse background components. For photo pion production Pierre Auger results imply a constrain value for the LIV energy scale, at $M \geq 50\ eV$.

Work partially supported by CONACYT México, under grant number 132197.

References

[1] Matteo Galverni, Günter Sigl 2008 *Phys. Rev. Lett.* **100** (2008) 021102.
[2] The Pierre Auger Collaboration 2010 *Phys. Lett.* B **685** 239-246 (2010).
[3] Mariangela Settimo for the Pierre Auger Collaboration 2011 Proceedings of 32th Int. Cosmic Ray Conf. (ICRC-2011).
[4] Roberto Aloisio, Pasquale Blasi, Piera L. Ghia, Aurelio Grillo 2000 *Phys. Rev.* D **62** 053010.
[5] T. Jacobson, S. Loerati, and D. Mattingly 2002 *Phys. Rev.* **D** 67 124011.
[6] Giovanni Amelino-Camelia 2012 *Phys. Rev.* **D** 85, 084034

Journal of Physics: Conference Series **468** (2013) 012006 doi:10.1088/1742-6596/468/1/012006

HIGGS BOSON DECAYS TO NEUTRALINOS AND CHARGINOS IN THE NMSSM

M. del R. Aparicio Méndez[(a)], **J. E. Barradas Guevara**[(a)], **L. A. Cordero Cid** [(b)] **and O. G. Félix Beltrán**[(b)]

[(a)]Facultad de Ciencias Físico Matemáticas, BUAP, Apdo. Postal 1152, 72570, Puebla, Puebla, México.
[(b)]Facultad de Ciencias de la Electrónica, Benemérita Universidad Autónoma de Puebla, Apdo. Postal 542, 72570, Puebla, Puebla, México.

E-mail: ma_rocio_a_m@yahoo.com.mx

Abstract. Within the framework of the Next-To-Minimal Supersymmetric Standard Model (NMSSM), we present the Feynman diagrams as well as the explicit formulas for the decay width of neutral Higgs boson to two charginos and two neutralinos. The importance of these decays is that they may have sizeable probability that the Higgs boson decays into LSPs.

1. Introduction

In this work we study the Next-to-Minimal Supersymmetric Standard Model (NMSSM) in which a singlet Higgs superfield is added to the two doublet superfields that are present in the minimal extension (MSSM) [1], [2].

The addition of a $SU(3) \times SU(2) \times U(1)$ gauge singlet supermultiplet to the MSSM has been motivated by several issues: (i) the solution of the so-called μ-problem; (ii) the possibility of spontaneous breaking of the CP symmetry; (iii) the upper limits on the Higgs mass that are relatively low in the MSSM could be relaxed; (iv) the neutral particle spectrum is enriched by two scalars and one fermion, which could mix with other Higgs bosons and neutralinos and modify their physical properties; (v) possible deviations from the MSSM could urge to enlarge the experimental searches, which have been centred upon the MSSM. We refer here to the supersymmetric model with a gauge singlet in its simplest version, the NMSSM [2], [4]. As compared to the MSSM, the NMSSM can induce a richer phenomenology in the Higgs and neutralino sectors.

2. Generalities

One of the most attractive supersymmetric models is the Next to Minimal Supersymmetric Standard Model [5] which extends the MSSM by the introduction of just one singlet superfield, $\widehat{S}$. When the scalar component of $\widehat{S}$ acquires a TeV scale vacuum expectation value (a very natural result in the context of the model), the superpotential term $\lambda \widehat{S}\widehat{H}_u\widehat{H}_d$ generates an effective $\mu\widehat{H}_u\widehat{H}_d$ interaction for the Higgs doublet superfields with $\mu = \lambda\langle S\rangle$. Such a term is essential for acceptable phenomenology. No other SUSY model generates this crucial component of the superpotential in as natural a fashion [6]. The additional singlet superfield of the NMSSM leads to extended Higgs and neutralino sectors.

The Higgs sector of the NMSSM consists of the usual two Higgs doublets H_u and H_d and an extra Higgs singlet $\widehat{S}$. The extra singlet field is allowed to couple only to the Higgs doublets of the model, the supersymmetrization of which is that the singlet field only couples to the Higgsino doublets. Consequently, the couplings of the singlet $\widehat{S}$ to gauge bosons and fermions will only be manifest via their mixing with the doublet Higgs fields. After the Higgs fields take on the VEV's and rotating away the Goldstone modes, we are left with a pair of charged Higgs bosons, 3 real scalar fields, and 2 pseudoscalar fields [7]. Therefore the Higgs sector of the NMSSM contains five physical neutral Higgs bosons, three Higgs scalars $S_a = h_a$ (with $a = 1, 2, 3$) and two pseudoscalars $P_\alpha = h_\alpha$ (with $\alpha = 1, 2$), and two degenerate physical charged Higgs particles $H^\pm$.

In contrast to the Higgs sector, the neutralino sector 2 is complemented only by the familiar SU(2) and U(1) gaugino mass terms, resulting in a much less complex parameter space. The extra singlet superfield adds an extra higgsino to the spectrum, often called a singlino, resulting in five neutralino states [8], [9].

With fixed parameters of the Higgs sector the masses and mixings of the neutralinos are determined by the two further parameters M and M' of the Lagrangian

$$\begin{aligned}\mathcal{L}_{m_{\tilde{\chi}_i^0}} &= \frac{1}{\sqrt{2}} ig\lambda^3(v_1\psi^1_{H_1} - v_2\psi^2_{H_2}) - \frac{1}{\sqrt{2}} ig^{'}\lambda'(v_1\psi^1_{H_1} - v_2\psi^2_{H_2}) \\ &-\frac{1}{2}M\lambda^3\lambda^3 - \frac{1}{2}M^{'}\lambda'\lambda' \\ &-\lambda x\psi^1_{H_1}\psi^2_{H_2} - \lambda v_1\psi^2_{H_2}\psi_N - \lambda v_2\psi^1_{H_1}\psi_N + kx\psi^2_N + h.c. \end{aligned} \quad (1)$$

In the basis

$$(\psi^0)^T = (-i\lambda_\gamma,\ -i\lambda_Z,\ \psi^a_H,\ \psi^b_H,\ \psi_N) \quad (2)$$

with

$$\begin{aligned}\psi^a_H &= \psi^1_{H_1}\cos\beta - \psi^2_{H_2}\,\mathrm{sen}\,\beta, \\ \psi^b_H &= \psi^1_{H_1}\,\mathrm{sen}\,\beta + \psi^2_{H_2}\cos\beta, \end{aligned} \quad (3)$$

the mass term of the Lagrangian reads like

$$\mathcal{L}_{m_{\tilde{\chi}_i^0}} = -\frac{1}{2}(\psi^0)^T\mathbf{Y}\psi^0 + H.c.. \quad (4)$$

The symmetric neutralino mixing matrix is given as

$$\mathbf{Y} = \begin{pmatrix} -Ms^2_W - M'c^2_W & (M'-M)s_Wc_W & 0 & 0 & 0 \\ (M'-M)s_Wc_W & -Mc^2_W - M's^2_W & m_Z & 0 & 0 \\ 0 & m_Z & -\lambda x\,\mathrm{sen}\,2\beta & \lambda x\cos 2\beta & 0 \\ 0 & 0 & \lambda x\cos 2\beta & \lambda x\,\mathrm{sen}\,2\beta & \lambda v \\ 0 & 0 & 0 & \lambda v & -2kx \end{pmatrix}.$$

It can be diagonalized by a unitary 5×5 matrix $\mathcal{N}$

$$\mathcal{N}\mathbf{Y}\mathcal{N}^{-1} = \begin{pmatrix} m_{\tilde{\chi}_1^0} & 0 & 0 & 0 & 0 \\ 0 & m_{\tilde{\chi}_2^0} & 0 & 0 & 0 \\ 0 & 0 & m_{\tilde{\chi}_3^0} & 0 & 0 \\ 0 & 0 & 0 & m_{\tilde{\chi}_4^0} & 0 \\ 0 & 0 & 0 & 0 & m_{\tilde{\chi}_5^0} \end{pmatrix} = \mathbf{m}_{\tilde{\chi}_i^0}, \quad (5)$$

with real and positive mass eigenvalues $\mathbf{m}_{\tilde{\chi}_i^0}$. If one tolerates negative eigenvalues, the diagonalization matrix can be chosen to be real. Then the absolute values of $m_{\tilde{\chi}_i^0} \gtrless 0$ are the physical neutralino masses. The upper 4×4 matrix of Y represents the neutralino mixing matrix of the MSSM with $\mu = \lambda x$ [10]. The remaining particle content is identical with that of the MSSM.

3. Decay widths

In this section we expose the *decay rate* Γ (the probability per unit time of carrying out a particular process [11]) and Feynman diagrams for the coupling of the neutral Higgs bosons to pairs of neutralinos and charginos in the NMSSM.

The formulas for the decay widths of a scalar or pseudoscalar Higgs boson into two neutralinos [12]:

$$\begin{aligned}\Gamma(h_a \to \tilde{\chi}_i^0 \tilde{\chi}_j^0) &= \frac{g^2 \lambda^{\frac{1}{2}}}{16\pi M_{h_a}^3 (1+\delta(i,j))} \left[\left(\frac{1}{g^2} \left[(Q_{aij}^{L''})^2 + (Q_{aji}^{L''})^2 \right] \right) \right. \\ &\quad \times \left(M_{h_a}^2 - m_{\tilde{\chi}_i^0}^2 - m_{\tilde{\chi}_j^0}^2 \right) - \left[4 m_{\tilde{\chi}_i^0} m_{\tilde{\chi}_j^0} \right] \\ &\quad \left. \times \left[\frac{1}{g^2} \left(Q_{aij}^{L''} \right) \left(Q_{aji}^{L''} \right) \right] \right] \end{aligned} \tag{6}$$

$$\begin{aligned}\Gamma(h_\alpha \to \tilde{\chi}_i^0 \tilde{\chi}_j^0) &= \frac{g^2 \lambda^{\frac{1}{2}}}{16\pi M_{h_\alpha}^3 (1+\delta(i,j))} \left[\left(\frac{1}{g^2} \left[(R_{\alpha ij}^{L''})^2 + (R_{\alpha ji}^{L''})^2 \right] \right) \right. \\ &\quad \times \left(M_{h_\alpha}^2 - m_{\tilde{\chi}_i^0}^2 - m_{\tilde{\chi}_j^0}^2 \right) + \left[4 m_{\tilde{\chi}_i^0} m_{\tilde{\chi}_j^0} \right] \\ &\quad \left. \times \left[\frac{1}{g^2} \left(R_{\alpha ij}^{L''} \right) \left(R_{\alpha ji}^{L''} \right) \right] \right] \end{aligned} \tag{7}$$

The decay of a scalar or pseudoscalar Higgs boson into two charginos

$$\begin{aligned}\Gamma(h_a \to \tilde{\chi}_i^+ \tilde{\chi}_j^-) &= \frac{g^2 \lambda^{\frac{1}{2}}}{16\pi M_{h_a}^3 (1+\delta(i,j))} \left[\left(\frac{1}{g^2} \left[(Q_{aij})^2 + (Q_{aji})^2 \right] \right) \right. \\ &\quad \times \left(M_{h_a}^2 - m_{\tilde{\chi}_i^+}^2 - m_{\tilde{\chi}_j^-}^2 \right) - \left[4 m_{\tilde{\chi}_i^+} m_{\tilde{\chi}_j^-} \right] \\ &\quad \left. \times \left[\frac{1}{g^2} \left(Q_{aij} \right) \left(Q_{aji} \right) \right] \right] \end{aligned} \tag{8}$$

$$\begin{aligned}\Gamma(h_\alpha \to \tilde{\chi}_i^+ \tilde{\chi}_j^-) &= \frac{g^2 \lambda^{\frac{1}{2}}}{16\pi M_{h_\alpha}^3 (1+\delta(i,j))} \left[\left(\frac{1}{g^2} \left[(R_{\alpha ij})^2 + (R_{\alpha ji})^2 \right] \right) \right. \\ &\quad \times \left(M_{h_\alpha}^2 - m_{\tilde{\chi}_i^+}^2 - m_{\tilde{\chi}_j^-}^2 \right) + \left[4 m_{\tilde{\chi}_i^+} m_{\tilde{\chi}_j^-} \right] \\ &\quad \left. \times \left[\frac{1}{g^2} \left(R_{\alpha ij} \right) \left(R_{\alpha ji} \right) \right] \right] \end{aligned} \tag{9}$$

In Figure 1, we can shown the Feynman diagrams for the following couplings: (i) when neutral scalar Higgs bosons decay into pairs of neutralinos the processes that occur are $\Gamma(h_a \to \tilde{\chi}_i^0 \tilde{\chi}_j^0)$,

(a = 1, 2, 3); and (ii) the neutral pseudoscalar Higgs bosons processes are $\Gamma(h_\alpha \to \widetilde{\chi}_i^0\widetilde{\chi}_j^0)$ ($\alpha = 1, 2$).

For the case in which the neutral scalar Higgs bosons decay into pairs charginos the processes presented are: (i) $\Gamma(h_a \to \widetilde{\chi}_i^+\widetilde{\chi}_j^-)$; and ($ii$) the neutral pseudoscalar Higgs bosons $\Gamma(h_\alpha \to \widetilde{\chi}_i^+\widetilde{\chi}_j^-)$.

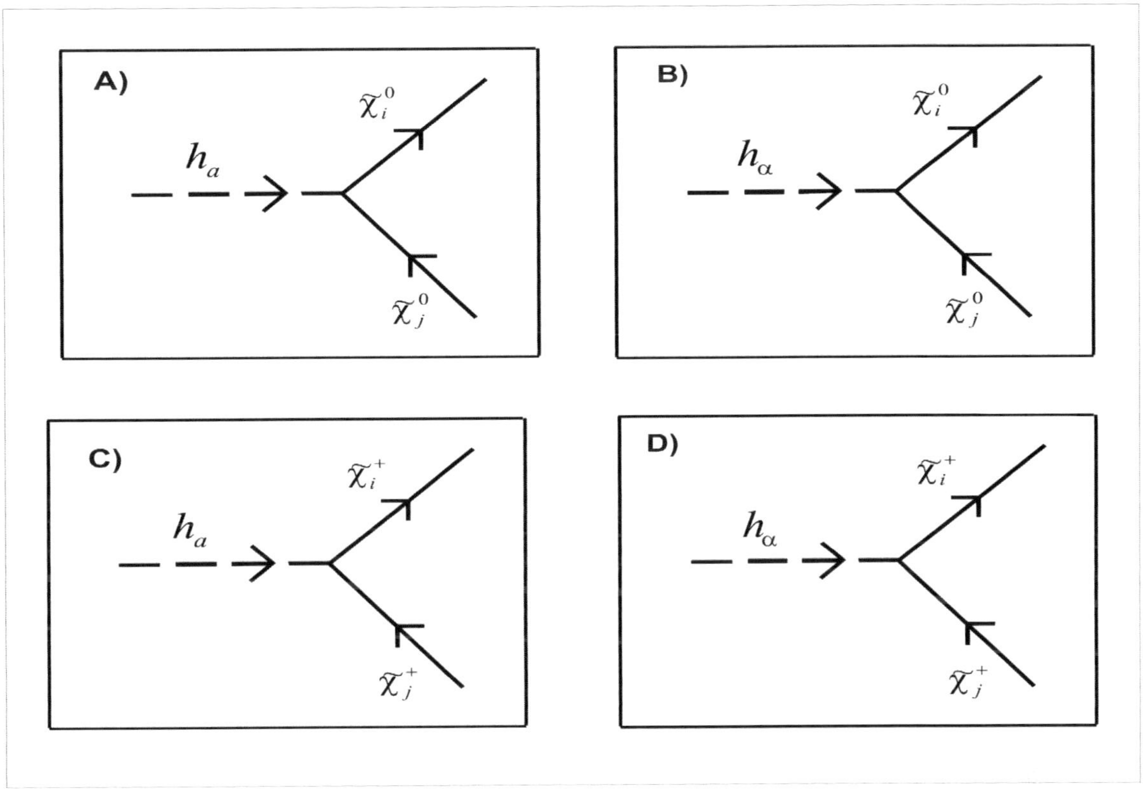

Figure 1. Feynman diagrams for the couplings of a scalar or pseudoscalar Higgs boson to two neutralinos ((A) and (B)) and two charginos ((C) and (D)).

4. Conclusion

The search for important particles such as Higgs bosons, leads to develop an exhaustive study on the possible decays in supersymmetric models like the NMSSM. In the framework of the basic supersymmetric model NMSSM, the decay width was calculated for the couplings of neutral Higgs bosons to charginos and neutralinos, recovering the equations decay widths incontrend in the literature for this kind of processes. A contribution of this work is to have obtained the formula that determines the decay width for neutral Higgs bosons in charginos.

5. References

[1] A. Djouadi *et al*, JHEP **0807**, 002 (2008) [arXiv:0801.4321].
[2] H. K. Dreiner, F. Staub and A. Vicente, Phys. Rev. D **87**, 035009 (2013) [arXiv:1211.6987].
[3] U. Ellwanger, M. Rausch de Traubenberg and C. A. Savoy, Nucl. Phys. B **492**, 21 (1997) [arXiv:9611251].
[4] S. F. King, M. Muhlleitner, R. Nevzorov and K. Walz, Nucl. Phys. B **870**, 323 (2013) [arXiv:1211.5074].
[5] J. F. Gunion and H. E. Haber, Nucl. Phys. B **272**, (1986).
[6] U. Ellwanger, J. F. Gunion and C. Hugonie, JHEP **0507**, 041 (2005) [arXiv:0503203].

Journal of Physics: Conference Series **468** (2013) 012006 doi:10.1088/1742-6596/468/1/012006

[7] K. Cheung and T. -J. Hou, Phys. Lett. B **674**, 54 (2009) [arXiv:0809.1122].
[8] D. J. Miller [arXiv:0410227].
[9] D. A. Vasquez, G. Belanger, C. Boehm, A. Pukhov and J. Silk, Phys. Rev. D **82**, 115027 (2010) [arXiv:1009.4380].
[10] F. Franke and H. Fraas, Int. J. Mod. Phys. A **12**, 479 (1997) [arXiv:9512366].
[11] David Griffiths, *Introduction to elementary particles*, Wiley, New York, 198.
[12] F. Franke and H. Fraas, Z. Phys. C **72**, 309 (1996) [arXiv:9511275].

Journal of Physics: Conference Series **468** (2013) 012007 doi:10.1088/1742-6596/468/1/012007

The $Z, Z' \to \gamma\gamma\gamma$ decays in the minimal 331 model

J Montaño[(a)], **M A Pérez**[(a)], **F Ramírez-Zavaleta**[(b)] **and J J Toscano**[(c)]
[(a)]Departamento de Física, CINVESTAV, Apartado Postal 14-740, 07000, México, D. F., México.
[(b)]Facultad de Ciencias Físico Matemáticas, Universidad Michoacana de San Nicolás de Hidalgo, Avenida Francisco J. Mújica S/N, 58060, Morelia, Michoacán, México.
[(c)]Facultad de Ciencias Físico Matemáticas, Benemérita Universidad Autónoma de Puebla, Apartado Postal 1152, Puebla, Puebla, México.

E-mail: `feramirez@umich.mx`

Abstract. A complete calculation at one-loop level for the $Z, Z' \to \gamma\gamma\gamma$ decays is presented in the context of the minimal 331 model, which predicts the existence of a new Z' neutral gauge boson, Y^{++}, Y^{+} charged gauge bosons, new exotic quarks and charged scalars. Bose symmetry is exploited to write a compact and manifest $SU_C(3)$-invariant vertex function for the $\gamma\gamma\gamma V^0$ coupling. Previous results on the $Z \to \gamma\gamma\gamma$ are reproduced ($\mathrm{Br}(Z \to \gamma\gamma\gamma) \sim 10^{-10}$). It is found that this decay is insensitive to the effects of new heavy particles. This contrasts with the $Z' \to \gamma\gamma\gamma$ decay where the associated branching ratio is of the order of 10^{-6}.

1. Introduction

In this work, we will study some properties of the Z' gauge boson predicted by the minimal 331 model [1]. Particularly, we are interested in studying the rare decay of this Z' gauge boson into three photons. This type of decay is naturally suppressed in renormalizable theories, as they first arise at the one-loop level. The analogous decay in the Standard Model (SM) $Z \to \gamma\gamma\gamma$ is very suppressed, with a branching ratio of the order of 10^{-10} GeV [2].

2. The minimal 331 model

The minimal 331 model is based on the $SU_C(3) \otimes SU_L(3) \otimes U_X(1)$ gauge group. The $SU_C(3)$ group represents strong interactions and the $SU_L(3) \otimes U_X(1)$ group is the simplest extension of the electroweak gauge group. The minimal 331 model predicts new exotic quarks, new gauge bosons and scalars, among them a new neutral gauge boson Z' is expected at the TeV scale [1]. This model offers a possible solution to the family replication problem: anomalies cancel out when all families are summed over, which suggests that the family number must be 3. The lepton spectrum of the model is accommodated as antitriplets of $SU_L(3)$:

$$L_{1,2,3} = \begin{pmatrix} e \\ \nu_e \\ e^c \end{pmatrix}, \begin{pmatrix} \mu \\ \nu_\mu \\ \mu^c \end{pmatrix}, \begin{pmatrix} \tau \\ \nu_\tau \\ \tau^c \end{pmatrix} : (1, 3^*, 0) .$$

Two quark generations are arranged as triplets and the other one as an antitriplet:

$$Q_{1,2} = \begin{pmatrix} u \\ d \\ D \end{pmatrix}, \begin{pmatrix} c \\ s \\ S \end{pmatrix} : (3, 3, -1/3);$$

$$Q_3 = \begin{pmatrix} b \\ t \\ T \end{pmatrix} : \quad (3, 3^*, 2/3);$$

The right-handed quarks are given as follows:

$$d^c,\ s^c,\ b^c : \ (3^*, 1, 1/3); \quad D^c,\ S^c : \ (3^*, 1, 4/3),$$
$$u^c,\ c^c,\ t^c : \ (3^*, 1, -2/3); \quad T^c : \ (3^*, 1, -5/3).$$

The D, S and T particles are new exotic quarks with charges $-4/3$, $-4/3$ and $+5/3$, respectively. Henceforth, we denote exotic quarks with Q. The scalar sector of the minimal 331 model is comprised of three triplets and one sextet of $SU_L(3)$:

$$\phi = \begin{pmatrix} \Phi_Y \\ \phi^0 \end{pmatrix} : (1, 3, 1), \quad \phi_1 = \begin{pmatrix} \Phi_1 \\ \delta^- \end{pmatrix} : (1, 3, 0),$$

$$\Phi_2 = \begin{pmatrix} \tilde{\Phi}_2 \\ \rho^{--} \end{pmatrix} : (1, 3, -1)$$

$$H = \begin{pmatrix} T & \tilde{\Phi}_3/\sqrt{2} \\ \tilde{\Phi}_3^T/\sqrt{2} & \eta'^0 \end{pmatrix} : \quad (1, 6, 0)\ .$$

The spontaneous symmetry breaking occurs in the following way: $SU_L(3) \otimes U_X(1) \xrightarrow{\phi_Y} SU_L(2) \otimes U_Y(1) \xrightarrow{\phi_1, \phi_2, H} U_e(1)$, where ϕ_Y gives masses to the new exotic quarks D , S , T and to the new heavy gauge bosons Z' , $Y^{\pm\pm}$, $Y^{\pm}$.

3. Vertex $\gamma\gamma\gamma V^0$, with $V \equiv Z, Z'$

The amplitude of the $\gamma\gamma\gamma V^0$ vertex can be written according to the spin of particles circulating in the loops (see Figure 1). This is possible since we calculate the gauge particles contribution using covariant R_ξ gauges [3], which render finite and gauge invariant contributions

$$\mathcal{M}_{\gamma\gamma\gamma V^0} = \mathcal{M}_{1/2} + \mathcal{M}_1 + \mathcal{M}_0,$$

where $\mathcal{M}_{1/2}$, $\mathcal{M}_1$ and $\mathcal{M}_0$ are, respectively, the spinorial, vectorial and scalar amplitudes.

In Fig. 1, X represents the type of particle circulating in the loops: fermions: u, d, s, c, b, t, D, S, T, e, μ, τ, gauge bosons: W^+, Y^+, Y^{++} auxiliar-pseudo scalar fields: G_W^+, C_W^+, $\bar{C}_W^+$, G_Y^+, C_Y^+, $\bar{C}_Y^+$, G_Y^{++}, C_Y^{++}, $\bar{C}_Y^{++}$ and scalars: h_1^+, h_2^+, h_3^+, h_4^+, d_1^{++}, d_2^{++}, d_3^{++} [3]. For each fermion: 6 boxes, for each gauge boson and scalar: 6 boxes, 12 triangles and 3 bubbles.

Once the loop integrals are computed, the amplitudes can be expressed in terms of gauge structures and their associated form factors as follows:

$$\mathcal{M}^{\mu_1\mu_2\mu_3\mu_4}_{\gamma\gamma\gamma V^0} = \frac{i}{\pi^2}\frac{ge^3}{2c_W}\sum_{i=1}^{18}\left(F^{\frac{1}{2}}_{V^0 i} + F^{1}_{V^0 i} + F^{0}_{V^0 i}\right)T_i^{\mu_1\mu_2\mu_3\mu_4}.$$

Bose statistics imposes that the amplitudes must be symmetric under the interchanges of pairs of photons: $(p_1, \mu_1) \leftrightarrow (p_2, \mu_2) \leftrightarrow (p_3, \mu_3)$. Gauge invariance is satisfied: $p_{i\mu_i}\mathcal{M}^{\mu_1\mu_2\mu_3\mu_4}_{\gamma\gamma\gamma V^0} = 0$. There are 18 Lorentz structures:

$$\begin{aligned}
T_{V1}^{\mu_1\mu_2\mu_3\mu_4} &= (p_{12}g^{\mu_1\mu_2} - p_2^{\mu_1}p_1^{\mu_2})(p_{13}g^{\mu_3\mu_4} - p_1^{\mu_3}p_3^{\mu_4}),\\
T_{V7}^{\mu_1\mu_2\mu_3\mu_4} &= (p_{13}p_2^{\mu_1} - p_{12}p_3^{\mu_1})(p_{23}g^{\mu_2\mu_3} - p_3^{\mu_2}p_2^{\mu_3})p_2^{\mu_4},\\
T_{V13}^{\mu_1\mu_2\mu_3\mu_4} &= (p_{13}g^{\mu_1\mu_2} - p_3^{\mu_1}p_1^{\mu_2})(p_{23}g^{\mu_3\mu_4} - p_2^{\mu_3}p_3^{\mu_4})\\
&\quad + (p_{12}p_3^{\mu_1} - p_{13}p_2^{\mu_1})(p_3^{\mu_2}g^{\mu_3\mu_4} - p_3^{\mu_4}g^{\mu_2\mu_3}).
\end{aligned}$$

where $p_{ij} \equiv p_i \cdot p_j$. The rest of structures are obtained from these ones by Bose symmetry. $F_{V_i^0}$ are finite scalar form factors in terms of Passarino-Veltman functions [3].

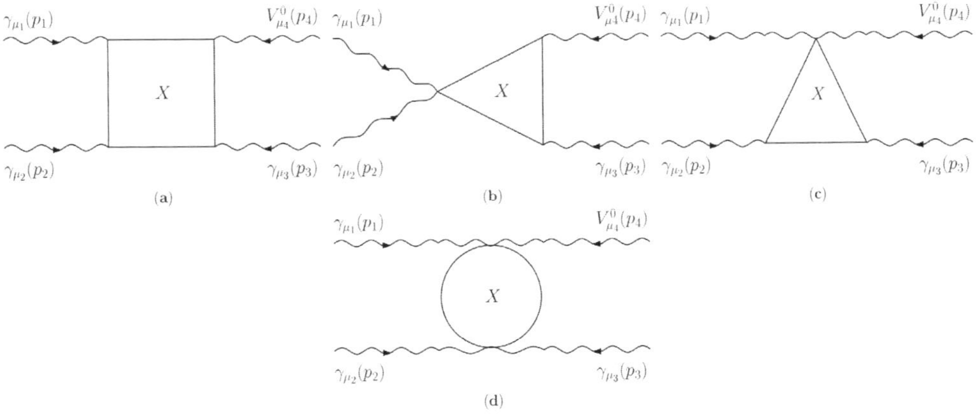

Figure 1. Feynman diagrams contributing to the $V^0 \to \gamma\gamma\gamma$ decay. Fermions only contribute through box diagrams as shown in (a).

4. The $Z \to \gamma\gamma\gamma$ and $Z' \to \gamma\gamma\gamma$ decays

By using the formula for calculating the decay width [3, 4], we can discuss the impact of new physics on the $Z \to \gamma\gamma\gamma$ decay. To perform this, we consider the scenario: $m_Q = 500$ GeV and $m_Y = m_H$ =250 GeV. The relative importance of each type of contribution is shown in Table 1. The Particle Data Group reports that $\Gamma(Z \to \gamma\gamma\gamma) < 10^{-5}$ GeV [4]. However, the SM prediction states that $\Gamma(Z \to \gamma\gamma\gamma) = 1.35 \times 10^{-9}$ GeV with an associated branching ratio of 5.41×10^{-10}. In contrast, the minimal 331 model prediction is $\Gamma(Z \to \gamma\gamma\gamma) = 1.31 \times 10^{-9}$ GeV with a respective branching ratio of 5.26×10^{-10}[3].

Sector	Br
Fermions	4.16×10^{-10}
Gauge Bosons	1.03×10^{-11}
Scalar	3.04×10^{-13}
Fermions-Gauge Bosons	9.92×10^{-11}
Fermions-Scalar	-5.90×10^{-14}
Gauge Bosons-Scalar	4.30×10^{-13}
Total	5.26×10^{-10}

Table 1. Contributions by spin of particle to the Br$(Z \to \gamma\gamma\gamma)$ in the scenario: $m_Q = 500$ GeV and $m_Y = m_H$ =250 GeV.

As referred to the minimal 331 model, it imposes the theoretical restriction $\frac{m_Y}{m_{Z'}} < 0.26$ [5], whereas lower bounds on m_Y and $m_{Z'}$ obtained from experimental data restrict this ratio to be $0.19 < \frac{m_Y}{m_{Z'}}$ [5]. The most promising scenario occurs when $m_Q = 500$ GeV, $\frac{m_Y}{m_{Z'}} = 0.19$, m_Y =275.5 GeV and $m_H = 250$ GeV. The total contribution to Br$(Z' \to \gamma\gamma\gamma)$ can be appreciated in Fig. 2. From this figure, we can see that the branching ratio for the $Z' \to \gamma\gamma\gamma$ is mainly of the order of 10^{-6}. For further details of calculation and analysis see Reference [3].

5. Final remarks

The minimal 331 model predicts new physic at energy relatively near to the Fermi scale and gives a possible solution to the fermion family replication problem. The $Z \to \gamma\gamma\gamma$ decay is insensitive to the presence of the new physics provided by the minimal 331 model. We corroborate that

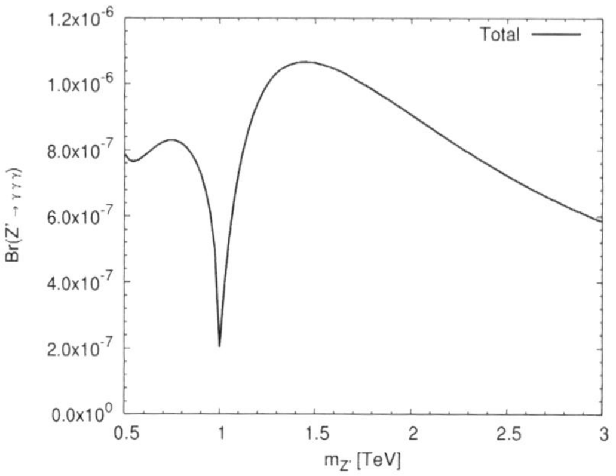

Figure 2. Total contribution to $\mathrm{Br}(Z' \to \gamma\gamma\gamma)$ as a function of $m_{Z'}$.

$\mathrm{Br}(Z \to \gamma\gamma\gamma) \sim 10^{-10}$. For the $Z' \to \gamma\gamma\gamma$ decay, the best signal is $\mathrm{Br}(Z' \to \gamma\gamma\gamma) \sim 1.07 \times 10^{-6}$ when $m_{Z'}$ =1.45 TeV in the scenario m_Q = 500 GeV, $m_Y/m_{Z'}$=0.19, m_Y =275.5 GeV and m_H = 250 GeV.

Acknowledgments

This work has been partially supported by SNI and CIC-UMSNH (México).

References

[1] F. Pisano and V. Pleitez, Phys. Rev. D **46**, 410 (1992); P. H. Frampton, Phys. Rev. Lett. **69**, 2889 (1992).
[2] V. N. Baier, E.A. Kurayev, and V. S. Fadin, Sov. J. Nucl. Phys. **31**, 364 (1980); M. L. Laursen, K. O. Mikaelian and M. A. Samuel, Phys. Rev. D **23**, 2795 (1981); J. J. van der Bij and E.W. N. Glover, Nucl. Phys. B **313**, 237 (1989); J. J. van der Bij and E.W. N. Glover, Report No. CERN 89-08, Vol. 2 p. 30; E.W. N. Glover and A. G. Morgan, Z. Phys. C **60**, 175 (1993); M. Baillargeon and F. Boudjema, Phys. Lett. B **272**, 158 (1991); X.Y. Pham, Phys. Lett. B **272**, 373 (1991); F. Dong, X. Jiang, and X. Zhou, Phys. Rev. D **47**, 5169 (1993); **46**, 5074 (1992).
[3] J. Montaño *et. al.*, Phys. Rev. D. **85**, 035012 (2012).
[4] J. Beringer *et. al.*, Phys. Rev. D **86**, 010001 (2012).
[5] P. H. Frampton, J. T. Liu, B. C. Rasco, and D. Ng, Mod. Phys. Lett. A **9**, 1975 (1994).

Journal of Physics: Conference Series **468** (2013) 012008 doi:10.1088/1742-6596/468/1/012008

QED effective action for an O(2)×O(3) symmetric field in the full mass range

N. Ahmadiniaz, A. Huet, A. Raya and C. Schubert

Instituto de Física y Matemáticas, Universidad Michoacana de San Nicolás de Hidalgo, Edificio C-3, Cd. Universitaria. C. P. 58040. Morelia, Michoacán, México.

E-mail: naser@ifm.umich.mx, raya@ifm.umich.mx, adolfo.huet@gmail.com, schubert@ifm.umich.mx

Abstract. An interesting class of background field configurations in QED are the $O(2) \times O(3)$ symmetric fields. Those backgrounds have some instanton-like properties and yield a one-loop effective action that is highly nontrivial but amenable to numerical calculation, for both scalar and spinor QED. Here we use the recently developed "partial-wave-cutoff method" for a numerical analysis of both effective actions in the full mass range. In particular, at large mass we are able to match the asymptotic behavior of the physically renormalized effective action against the leading two mass levels of the inverse mass (or heat kernel) expansion. At small mass we obtain good numerical results even in the massless case for the appropriately (unphysically) renormalized effective action after the removal of the chiral anomaly term through a small radial cutoff factor. In particular, we show that the effective action after this removal remains finite in the massless limit, which also provides indirect support for M. Fry's hypothesis that the QED effective action in this limit is dominated by the chiral anomaly term.

1. The $O(2) \times O(3)$ symmetric background

G. Dunne *et al* [1, 2, 3, 4] initiated the application of the "partial wave-cutoff method", to be explained below, to the important class of $O(2) \times O(3)$ symmetric fields first introduced by S. L. Adler [5, 6]. These backgrounds can be defined, in Euclidean metric, as

$$A_\mu(x) = \eta^3_{\mu\nu} x_\nu g(r) \, , \quad g(r) \equiv \nu \, \frac{e^{-\alpha r^2}}{\rho^2 + r^2} \, , \tag{1}$$

where $\eta^3_{\mu\nu}$ is a 't Hooft symbol, $r^2 = x_\mu x^\mu$ and $\alpha \geq 0$. Note that $g(r)$ is spherically symmetric, and has the further properties

(i) $\alpha > 0 \rightarrow \int d^4x \, F^2$ is finite.

(ii) $\alpha = 0 \rightarrow g(r) \propto \frac{1}{r^2}$, which is what we need to see the chiral anomaly term $\int d^4x F_{\mu\nu}\tilde{F}_{\mu\nu}$.

According to M. Fry [7, 8], the following general remarks hold for the spinor QED effective action in the background (1) with $\alpha = 0$: let $\mathcal{R}$ denote the (scheme independent) effective action obtained after subtraction of the two-point contribution. It behaves for small m as

$$\mathcal{R} \sim \frac{\nu^2}{4} \ln m^2 + \text{less singular in } m^2 \, . \tag{2}$$

The logarithmic term is determined entirely by the chiral anomaly,

$$-\frac{1}{4\pi^2}\int d^4x F_{\mu\nu}\tilde{F}_{\mu\nu}=\frac{\nu^2}{2}. \tag{3}$$

2. The partial wave-cutoff method

After decomposing the negative chirality part of the Dirac operator into partial-wave radial operators with quantum numbers l and l_3, the corresponding effective action is :

$$\Gamma_{\rm L}^{(-)}=-\sum_{s=\pm\frac{1}{2}}\sum_{l=0,\frac{1}{2},1,\ldots}^{L}\Omega(l)\sum_{l_3=-l}^{l}\ln\left(\frac{\det(m^2+\mathcal{H}_{(l,l_3,s)})}{\det(m^2+\mathcal{H}^{\rm free}_{(l,l_3,s)})}\right). \tag{4}$$

We concentrate on the negative chirality sector of the spinor effective action where $\Omega(l)=(2l+1)$ is the degeneracy factor, and the s sum comes from adding the contributions of each spinor component. The partial-wave cutoff method separates the sum over the quantum number l into a low partial-wave contribution, each term of which is computed using the (numerical) Gel'fand-Yaglom method, and a high partial-wave contribution, whose sum is computed analytically using WKB. Then we apply a regularization and renormalization procedure and combine these two contributions to yield the finite and renormalized effective action. The Gel'fand-Yaglom method [1, 2, 3], can be summarized as follows: Let $\mathcal{M}_1$ and $\mathcal{M}_2$ denote two second-order radial differential operators on the interval $r\in[0,\infty)$ and let $\Phi_1(r)$ and $\Phi_2(r)$ be solutions to the initial value problem

$$\mathcal{M}_i\Phi_i(r)=0;\quad \Phi_i(r)\sim r^{2l}\quad \text{as}\quad r\to 0. \tag{5}$$

Then the ratio of the determinants is given by

$$\frac{\det\mathcal{M}_1}{\det\mathcal{M}_2}=\lim_{R\to\infty}\left(\frac{\Phi_1(R)}{\Phi_2(R)}\right).$$

In our case

$$\Phi_-''(r)+\frac{4l+3}{r}\Phi_-'-\left(m^2+4l_3g(r)+r^2g(r)^2+[4g(r)+rg'(r)]\right)\Phi_-(r)=0.$$

The high-mode contribution, which remains to be calculated calculated using WKB, is

$$\Gamma_{\rm H}^{(-)}=-\sum_{s=\pm\frac{1}{2}}\sum_{l=L+\frac{1}{2}}^{\infty}\Omega(l)\sum_{l_3=-l}^{l}\ln\left(\frac{\det(m^2+\mathcal{H}_{(l,l_3,s)})}{\det(m^2+\mathcal{H}^{\rm free}_{(l,l_3,s)})}\right). \tag{6}$$

3. Two versions of the effective action

For the class of backgrounds considered here, the partial-wave-cutoff method works well for any value of the mass up to numerical accuracy. The effective action calculated as above is finite for any non-zero value of the mass. When we use on-shell ('OS') renormalization ($\mu=m$), its leading small-mass behavior contains the logarithmically divergent term [4]

$$\Gamma_{\rm ren}^{\rm OS}(m)\sim\left(-\int_0^\infty dr\,Q_{\log}(r)\right)\ln m,\qquad m\to 0. \tag{7}$$

Thus for the study of this small m regime we introduce a modified effective action,

$$\tilde{\Gamma}_{\rm ren}(m)\equiv\Gamma_{\rm ren}(m,\mu)+\left(\int_0^\infty dr\,Q_{\log}(r)\right)\ln\mu\quad\Big(=\Gamma_{\rm ren}(m,\mu=1)\Big). \tag{8}$$

It turns out that $\tilde{\Gamma}$ is finite for $m=0$, which supports Fry's conjecture, mentioned above, for the case of the backgrounds with $\alpha>0$ (where the chiral anomaly term is absent). Both variants of the effective action are contrasted in Fig. 1.

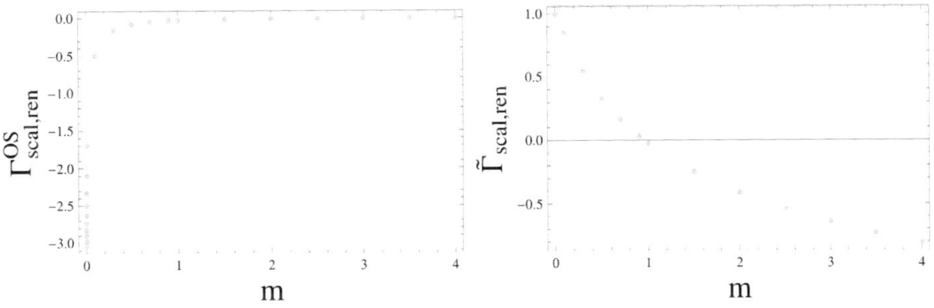

Figure 1. Effective action. *Left panel*, On shell, eq. (7). *Right panel*, modified, eq. (8).

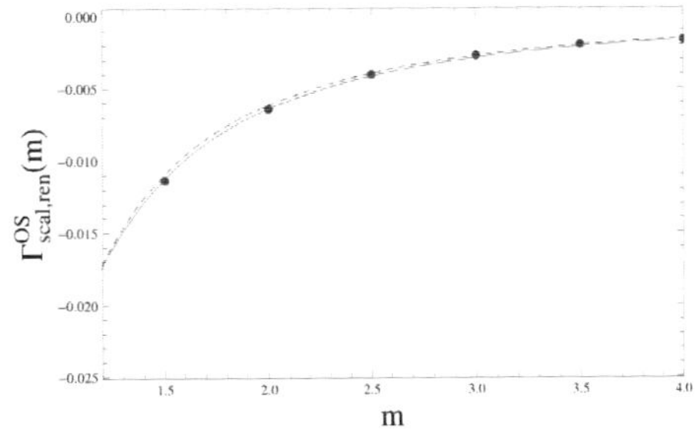

Figure 2. Large mass limit of the scalar effective action for $\alpha = 1/120$. Dots represent the exact effective action, the dashed curve considers only the leading term c_2/m^2, and the solid curve, both the leading and subleading terms $c_2/m^2 + c_4/m^4$.

4. Large mass asymptotic behavior

In this section we exhibit the leading and subleading terms in the inverse mass (= heat kernel) expansion of the one-loop scalar QED effective action. The first two terms are (we calculated them using the worldline formalism along the lines of [9, 10])

$$\Gamma^{\rm OS}_{\rm scal}(m) = \frac{c_{\rm scal,2}(\alpha)}{m^2} + \frac{c_{\rm scal,4}(\alpha)}{m^4} + O\left(\frac{1}{m^6}\right), \tag{9}$$

where the coefficients in the limit $\alpha \to 0$ are, up to cubic order in α,

$$\begin{aligned} c_{\rm scal,2}(\alpha) &= -\frac{1}{460}\alpha^3\left(256\log(2\alpha) + 256\gamma_E + \frac{304}{5}\right) - \frac{31\alpha^2}{300} + \frac{23\alpha}{600} - \frac{2}{75}, \\ c_{\rm scal,4}(\alpha) &= -\frac{37\gamma_E\alpha^3}{135} + \frac{187349\alpha^3}{396900} - \frac{2\gamma_E\alpha^2}{15} - \frac{36853\alpha^2}{529200} - \frac{1}{135}(37\alpha+18)\alpha^2\log(2\alpha) \\ &\quad - \frac{571\alpha}{22050} + \frac{107}{52920}. \end{aligned} \tag{10}$$

The large-mass behavior of the effective action is shown in Fig. 2 for the scalar QED case.

5. Finiteness of the massless four-point contribution

In this section we show that the four-point contribution to the effective action in the "standard" $O(2) \times O(3)$ symmetric background, (1) with $\alpha = 0$ and $\nu = \rho = 1$, is finite in the massless limit. This is a detail of some importance for Fry's investigation that had been missing in the analysis of [7], although it has been anticipated in [8]. In the worldline formalism, we can write this quartic contribution to the effective action as (in either scalar or spinor QED)

$$\Gamma^{(4)}[A] = -\prod_{i=1}^{4}\int \frac{d^4k_i}{(2\pi)^4}\bar{a}(k_i^2)(2\pi)^4\delta^4(\sum k_i)\Gamma[k_1,\varepsilon_1;\cdots;k_4,\varepsilon_4] \,, \tag{11}$$

where Γ is the worldline path integral representation of the off-shell Euclidean four-photon amplitude and $\bar{a}(k^2) = 4c\pi^2 K_2(\rho\sqrt{k^2})/k^2$, where $K_2(x)$ is the modified Bessel function of the second kind. After performing the path integral, suitable integrations by parts, a rescaling $\tau_i = Tu_i, i = 1,\ldots,4$ and the elimination of the global T integral, we obtain (see [10] for details)

$$\Gamma[k_1,\varepsilon_1;\cdots;k_4,\varepsilon_4] = -\frac{e^4}{(4\pi)^2}\int_0^1 du_1du_2du_3du_4\,\frac{Q_4(\dot{G}_{B12},\ldots,\dot{G}_{B34})}{\left(m^2-\frac{1}{2}\sum_{i,j=1}^4 G_{Bij}k_i\cdot k_j\right)^2}\,. \tag{12}$$

Here $G_{Bij} \equiv G_B(u_i,u_j) = |u_i-u_j| - (u_i-u_j)^2$ is the worldline Green's function and $\dot{G}_{Bij}$ its derivative. Q_4 is a polynomial in the various $\dot{G}_{Bij}$'s, as well as in the momenta and polarizations. Now, the QED Ward identity implies that the rhs of (12) is $O(k_i)$ in each of the four momenta, which can also be easily verified using properties of the numerator polynomial Q_4. Using this fact and (11) we see that there is no singularity at $k_i = 0$, and convergence at large k_i.

6. Conclusions

We have continued and extended here the full mass range analysis of the scalar and spinor QED effective actions for the $O(2)\times O(3)$ symmetric backgrounds, started in [4], by a more detailed numerical study of both the small and large mass behaviors. In [4] only the unphysically renormalized versions $\tilde{\Gamma}_{\rm ren}(m)$ of these effective actions were considered (corresponding to $\mu = 1$), which are appropriate for the small mass limit, but have a logarithmic divergence in m in the large m limit. Here we have instead used the physically renormalized effective actions $\Gamma_{\rm ren}^{\rm OS}(m)$ for the study of the large mass expansions, which made it possible to achieve a numerical matching of both this leading and even the subleading term in the inverse mass expansions of the effective actions. In our study of the small mass limit, we have improved on [4] by obtaining good numerical results for $\tilde{\Gamma}_{\rm ren}(m)$ even at $m = 0$, and showing continuity for $m \to 0$ for various values of α. Moreover, we have presented numerical evidence that $\tilde{\Gamma}_{\rm ren}(m = 0)$ stays finite even in the limit $\alpha \to 0$. This fact is important in the spinor case, where it supports indirectly Fry's conjecture [7] that, for the case at hand, the only source of a divergence of $\tilde{\Gamma}_{\rm ren}(m)$ for $\alpha = 0$ at $m \to 0$ should be the chiral anomaly term. As a side result, we have proved the finiteness of the massless limit four-point contribution to the effective action in scalar and spinor QED for the standard $O(2)\times O(3)$ symmetric background ($\alpha = 0$, $\nu = \rho = 1$).

References

[1] G. V. Dunne, J. Hur, C. Lee and H. Min, Phys. Rev. Lett. **94**, 072001 (2005), arXiv:hep-th/0410190.
[2] G. V. Dunne, J. Hur, C. Lee and H. Min, Phys. Rev. D **71**, 085019 (2005), arXiv:hep-th/0502087.
[3] G. V. Dunne, J. Hur and C. Lee, Phys. Rev. D **74**, 085025 (2006), arXiv:hep-th/0609118.
[4] G. V. Dunne, A. Huet, J. Hur and H. Min, Phys. Rev. D **83**, 105013 (2011).
[5] S. L. Adler, Phys. Rev. D **6**, 3445 (1972); Erratum-ibid. D **7**, 3821 (1973).
[6] S. L. Adler, Phys. Rev. D **10**, 2399 (1974); Erratum-ibid. D **15**, 1803 (1977).
[7] M. P Fry, Phys. Rev. D **75**, 065002 (2007), hep-th/0612218; Erratum-ibid. D **75** 069902 (2007).
[8] M. P Fry, Phys. Rev. D **81**, 107701 (2010).
[9] M.G. Schmidt and C. Schubert, Phys. Lett. B **318**, 438 (1993), hep-th/9309055.
[10] C. Schubert, Phys. Rept. **355**, 73 (2001), arXiv:hep-th/0101036.

Journal of Physics: Conference Series **468** (2013) 012009 doi:10.1088/1742-6596/468/1/012009

Solution of the Quark Gap Equation by using Gluon Propagator Models inspired by Lattice QCD

A Ahmad[1,2], **A Raya**[1]

[1]Instituto de Física y Matemáticas, UMSNH México.

[2] Department of Physics, Gomal University, Pakistan.

E-mail: (aftabahmad@ifm.umich.mx, raya@ifm.umich.mx)

Abstract. In this work we present the solution of the Schwinger-Dyson equation (SDE) for the quark propagator with gluon propagator models inspired by Lattice Quantum Chromodynamics (QCD) and a model of running coupling. Mass function $M(p^2)$ of the quarks and wave function renormalization $F(p^2)$ are obtained for the different models respectively. We also calculated the chiral quark condensate for each model and test for confinement through violation of positivity of the dynamical fermion propagator.

1. Gap equation

In this proceeding we explore the QCD gap equation

$$S(p)^{-1} = \imath\gamma \cdot p + m_d + \Sigma(p) \tag{1}$$

where

$$\Sigma(p) = \int \frac{d^4k}{(2\pi)^4} g^2 D_{\mu\nu}(k-p) \frac{\lambda^a}{2} \gamma^\mu S(k) \frac{\lambda^a}{2} \Gamma_\nu(k,p). \tag{2}$$

Our truncation is defined such that $\Gamma_\nu(k,p) = \gamma_\nu$, $g^2 D_{\mu\nu}(k-p) = D(q^2)(\delta_{\mu\nu} - q_\mu q_\nu/q^2)$ is the dressed gluon propagator, which is modeled or takes a form given by solving the coupled equation between gluon and ghost or by parameterizing the data of lattice QCD simulations, m_d is the current quark mass and λ^a are the Gell-Mann's matrices of the color group $SU(3)_c$. The quark propagator $S(k)$ is decomposed as

$$S(k) = \frac{F(k^2)}{\imath\gamma \cdot k + M(k^2)} = \imath\gamma \cdot k\sigma_v(k^2) + \sigma_s(k^2), \tag{3}$$

where $F(k^2)$ and $M(k^2)$ are the renormalized wave function and mass function of the quark respectively, and

$$\sigma_s(k^2) = \frac{F(k^2)M(k^2)}{k^2 + M^2(k^2)}, \qquad \sigma_v(k^2) = \frac{F(k^2)}{k^2 + M^2(k^2)}. \tag{4}$$

The order parameter of the dynamical chiral symmetry breaking is the chiral condensate,

$$-\langle \bar{q}q \rangle = \frac{3}{4\pi^2} \int_0^{\Lambda^2} dp^2 \frac{p^2 F(p^2) M(p^2)}{p^2 + M(p^2)^2}, \tag{5}$$

which we explore below for different models of the gap kernel. From Eq.(2) we have

$$\begin{aligned}\frac{1}{F(p^2)} &= 1+\frac{1}{3\pi^3 p^2}\int d^4k\mathcal{D}(q^2)\sigma_v(k^2)\left[(k\cdot p)+\frac{2(k\cdot q)(p\cdot q)}{q^2}\right], \\ \frac{M(p^2)}{F(p^2)} &= m_q+\frac{1}{\pi^3}\int d^4k\mathcal{D}(q^2)\sigma_s(k^2)\,. \end{aligned} \tag{6}$$

We take $\mathcal{D}(q^2) = A_2\alpha(q^2)D(q^2)$, and consider the model for the strong coupling [1]

$$\alpha_s(q^2) = \frac{a+b\frac{q^2}{\Lambda^2}}{1+c\frac{q^2}{\Lambda^2}+d\frac{q^4}{\Lambda^4}+o\frac{q^6}{\Lambda^6}}+\frac{\pi\gamma}{\log\left(e+\frac{q^2}{\Lambda^2}\right)} \tag{7}$$

where $a = 1.47$; $b = 0.881$; $c = 0.314$; $d = 0.00986$; $o = 0.00168$; $\gamma = 12/25$; $\Lambda = 0.234$, all quantities in the appropriate units of GeV, and $D(q^2)$ is the model of the gluon propagator obtained from lattice simulations. We have surveyed different types of models found in the literature [2] and select the best suitable models from Lattice QCD.

- The Lienweber model (LINA) [3] is of the form:

$$D(q^2) = Z\left\{\frac{A\Omega^{2\alpha}}{(q^2+\Omega^2)^{1+\alpha}}+\frac{L(q^2,\Omega)^2}{q^2+\Omega^2}\right\} \tag{8}$$

  where $Z = 2.01$, $\Omega = 0.5$, $A = 9.84$, $\alpha = 2.17$. Here

$$L(q^2,\Omega^2) = [0.5\log(q^2+\Omega^2)(q^{-2}+\Omega^{-2})]^{-\frac{13}{22}} \tag{9}$$

  is the infrared regulated version of the one loop logarithmic correction in ordered to ensures that this model have a correct leading ultraviolet behavior.

- The Refined Gribov-Zwanziger(RGZ) is of the form [4]:

$$D(q^2) = \frac{q^2+M_2}{q^4+\mu_2 q^2+\lambda_4} \tag{10}$$

  where $M_2 = 2.15$; $\mu_2 = 0.34$; $\lambda_4 = 0.2685$.

For comparison, we consider the well known Maris-Tandy model (MT) of the gap kernel [5]

$$4\pi\alpha\ D(q^2) = \frac{4\pi^2 Dq^2\exp(-\frac{q^2}{\Omega^2})}{\Omega^6}+\frac{8\pi^2\gamma\left(1-\exp\{-\frac{q^2}{4m_t^2}\}\right)}{q^2\log\left(\tau+(1+\frac{q^2}{\Lambda^2})^2\right)} \tag{11}$$

where $m_t = 0.5GeV$; $\gamma = 12/(33-2N_f)$; $N_f = 4$; $\tau = e^2-1$; $D = (0.96)^2$; $\Omega = 0.4$. This model has been widely used in SDE studies of hadron phenomenology. Finally, the A_2 parameter is selected such that the height of the mass functions for the three models are the same. The three models are compared in Fig. 1.

2. Numerical results

The mass function and wave function renormalization for the lattice inspired gluon models are shown in Figs. 2 and 3. For comparison, the results for the chiral condensate are tabulated in Table 1 for different current quark masses.

For the quark confinement test, we use the spatially averaged Schwinger function

$$\Delta(T) = \frac{1}{\pi}\int_0^\infty dp\cos(pT)\frac{F(p^2)M(p^2)}{p^2+M(p^2)^2}\,. \tag{12}$$

We check for positiveness of this function. Logarithm of $\Delta(T)$ for various values of the current quark mass are shown in Fig. 4. The dips in these curves signal confinement.

Journal of Physics: Conference Series **468** (2013) 012009 doi:10.1088/1742-6596/468/1/012009

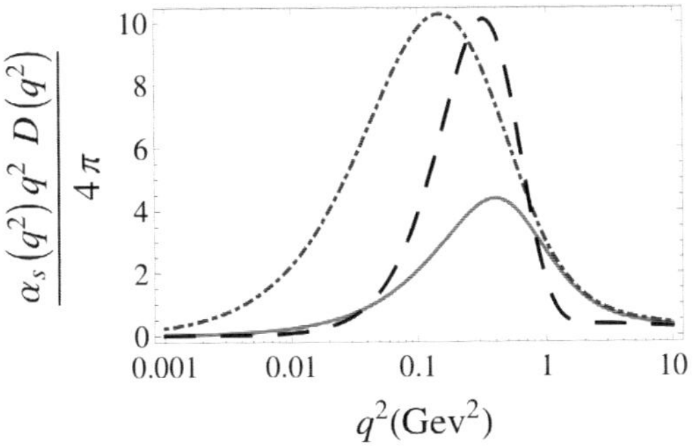

Figure 1. The gluon dressing function: The solid curve represents the RGZ model, the dashed curve represent the MT model and the dotted dashed curve represent the LINA model.

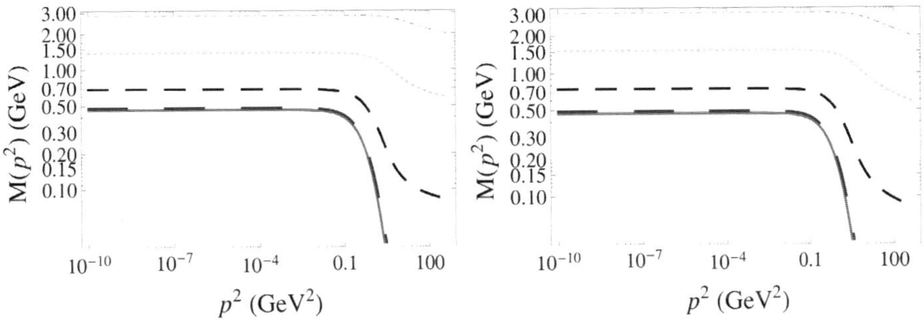

Figure 2. Mass function. *Left panel:* LINA model. *Right panel:* RGZ model. Solid curves represent the chiral limit $m_q = 0$GeV, long dashed curve represent $m_q = 0.0037$GeV, short dashed curve $m_q = 0.082$GeV, dotted curve $m_q = 0.59$GeV and dotted-dashed curve $m_q = 2.0$GeV.

Current Quark Masses(GeV)	MT	LINA	RGZ
$m_q = 0$	0.278536	0.281368	0.343
$m_{u/d} = 0.0037$	0.537689	0.544319	0.563984
$m_s = 0.082$	1.45734	1.45496	1.47338
$m_c = 0.59$	2.793	2.78651	2.81001
$m_b = 2.0$	4.06249	4.11062	4.13086

Table 1. Chiral condensate (in units of GeV^3) for different truncations of the gap equation.

3. Discussion and Conclusions

In this work we have presented the solution of QCD gap equation with models for gluon propagator inspired by Lattice QCD. The Lienweber model and Refined Gribov Zwinzager (RGZ) model are compared against the MT model which is known for last fifteen years in the SDE literature. The Lienweber and RGZ models with coupling model and with an appropriate weighting parameter A_2 parameter (A_2=1.3 for RGZ and A_2=0.8 for LinA model) yield good agreement with the well established MT results. All the models have shown enhancement in the

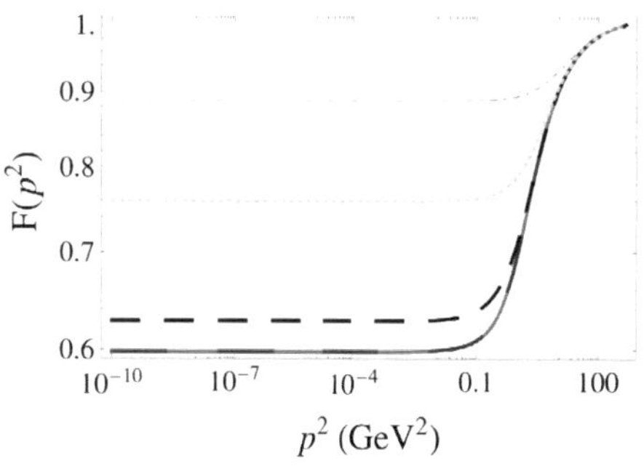

Figure 3. Wave function renormalization. *Left panel:* LINA model. *Right panel:* RGZ model. Solid curves represent the chiral limit $m_q = 0\text{GeV}$, long dashed curve represent $m_q = 0.0037\text{GeV}$, short dashed curve $m_q = 0.082\text{GeV}$, dotted curve $m_q = 0.59\text{GeV}$ and dotted-dashed curve $m_q = 2.0\text{GeV}$.

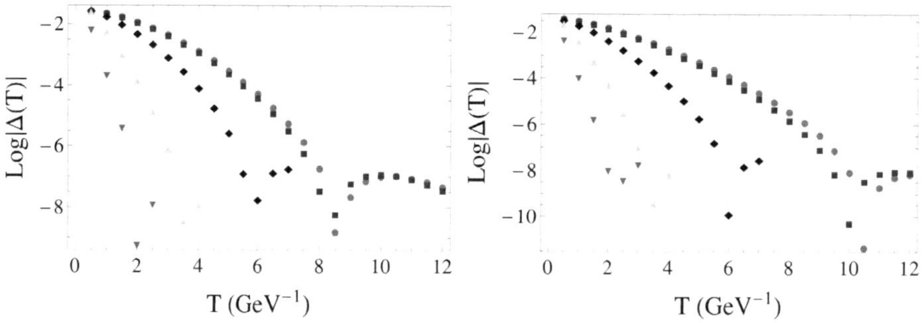

Figure 4. Confinement test. *Left panel:* LINA model. *Right panel:* RGZ model. Circles correspond to the chiral limit $m_q = 0\text{GeV}$, squares $m_q = 0.0037\text{GeV}$, diamonds $m_q = 0.082\text{GeV}$, upper-triangles $m_q = 0.59\text{GeV}$ and lower-triangles $m_q = 2.0\text{GeV}$.

infrared for the mass function and falling down to the ultraviolet. Wave function renormalization for the quark approaches to one for each model in the ultraviolet. All the three truncations give results for the condensate in good agreement with accepted phenomenological values of this quantity. All the solutions show neat signals of confinement.

Acknowledgments

We acknowledge CONACYT and CIC-UMSNH for support. We also acknowledge to Alejandro Ayala, Adnan Bashir and Axel Weber for valuable discussions. AA acknowledges the organizing committee for financial support.

References

[1] Maris P, Raya A, Roberts C D and Schmidt S M 2003, *Eur. Phys. J. A*18 231.
[2] Ahmad A 2006, *"Non-perturbative Gluon Propagator"* M.Phil Thesis, CHEP University of the Punjab, Lahore, Pakistan.
[3] Lienweber D, Skullerd J I, Williams A G and Parrinello C 1999, *Phys. Rev. D* **60** 094507.
[4] Dudal D, Oliveira O and Vandersickel N 2010, *Phys. Rev. D* **81** 074505.
[5] Maris P , Tandy P C 1999, *Phys. Rev. C***60** 055214.

Journal of Physics: Conference Series **468** (2013) 012010 doi:10.1088/1742-6596/468/1/012010

Dynamical Masses and Confinement in Graphene

S Hernández-Ortíz[1] and A Raya[1]

[1]Instituto de Física y Matemáticas, UMSNH.

E-mail: sortiz@ifm.umich.mx, raya@ifm.umich.mx

Abstract. Dynamical Mass Generation and Confinement are non-perturbative phenomena with large impact on the study of hadron and condensed matter physics. We study these phenomena in the light of the Schwinger-Dyson equations in various field theories, like Quantum Electrodynamics (QED) and Quantum Electrodynamics in the plane (QED3). Our aim is to describe the impact of these phenomena for the semi-metal–insulator transition in graphene, which is modeled by a peculiar field theory that inherits properties of both QED and QED3.

1. Introduction

Graphene, is a novel material which consists of a single layer of carbon atoms packed in a hexagonal (honeycomb) lattice. The low energy excitations near the Dirac points K and K' of the first Brillouin zone yields a gapless linear spectrum [1], and thus correspond to ultrarelativistic fermions or "graphinos" [2]. Graphene differs from most conventional three-dimensional materials. Intrinsic graphene is a semi-metal or zero-gap semiconductor. Understanding its electronic structure is the starting point for finding the band structure of graphite. In this work, we study the gap equation for graphinos and its similarities to the gap equation for electrons in QED3 and QED4 in the Rainbow Approximation. This is the basis of the work [3].

2. Rainbow Approximation

Schwinger-Dyson equations (SDEs) are general relations among the Green functions of a given Quantum Field Theory. These correspond to Euler-Lagrange equations of quantum field theories, since they are the equations of motion for Green functions. In QED, the SDE for the electron propagator is coupled to that of the photon propagator and the electron-photon vertex. Therefore, in order to find the electron propagator self-consistently, one must assume a particular form of the photon propagator and electron-photon vertex. In this work we are interested in the *Rainbow Approximation*, in which we consider the tree level contribution for both these Green functions. The diagram for the SDE in this approximation is

and corresponds to the expression

$$S^{-1}(p) = S_0^{-1}(p) - ie^2 \int \frac{d^d k}{(2\pi)^d} \gamma^\mu S(k) \gamma^\nu \Delta^0_{\mu\nu}(q) \,. \tag{1}$$

Journal of Physics: Conference Series **468** (2013) 012010 doi:10.1088/1742-6596/468/1/012010

Here, e^2 is the electromagnetic coupling, which in dimensions other than 4 is dimensionful, and $q = k - p$. The bare photon propagator is

$$\Delta^0_{\mu\nu}(q) = \frac{1}{q^2}\left(g_{\mu\nu} + (\xi - 1)\frac{q_\mu q_\nu}{q^2}\right),$$

and $S(p)$ is the electron propagator, which we prefer to write in its most general form as

$$S(p) = \frac{F(p^2)}{\not{p} - M^2(p^2)} = \sigma_v(p^2)\not{p} + \sigma_s(p^2) \quad (2)$$

where $F(p^2)$ and $M(p^2)$ are the electron wavefunction renormalization and mass function. The bare propagator $S_0(p)$ corresponds to $F(p^2) = 1$ and $M(p^2) = m$. Equivalently, $\sigma_v(p^2)$ and $\sigma_s(p^2)$ are the vector and scalar parts of the propagator, respectively.

A standard procedure reveals that in the Landau gauge ($\xi = 0$), $F(p^2) = 1$, and thus the gap equation of QED in arbitrary space-time dimensions reads [3]

$$M(p^2) = m - ie^2 \int \frac{d^dk}{(2\pi)^d}\frac{(d-1)M(k^2)}{q^2(k^2 + M^2(k^2))}, \quad (3)$$

where m is the electron bare mass, which we set to zero from now onward. $M(p^2)$ is the electron mass function. A non trivial solution $M(p^2) \neq 0$ to this expression corresponds to chiral symmetry breaking due to self-interactions. This means that we can have electron mass even if we started with $m = 0$. Below we discuss the nontrivial solutions to the above equation in 3 and 4 space-time dimensions and their role in the description of confinement.

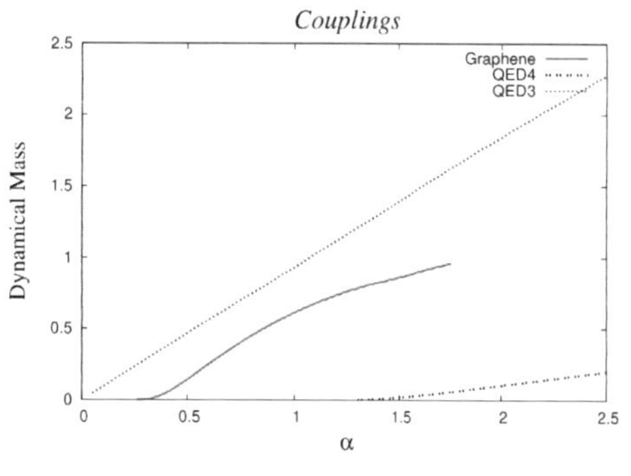

Figure 1. Scaling Law for QED, QED3 and Graphene.

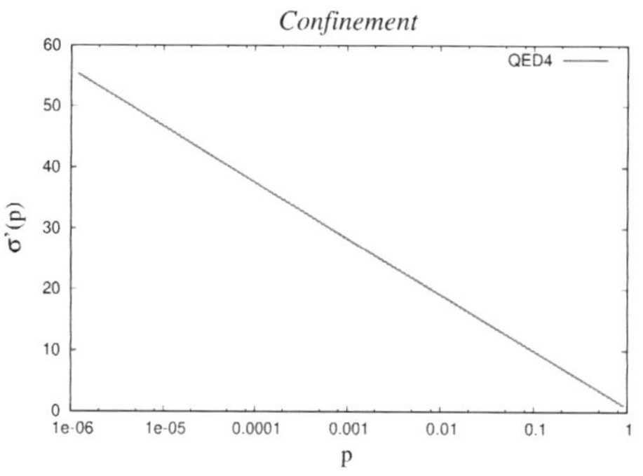

Figure 2. Confinement Test QED.

3. Gap equation in QED

In this section we work in the 4 dimensional space-time of ordinary QED. In order to solve Eq.(1), we perform a Wick rotation ($p^0 \to ip^0$) to Euclidean space. Then, the gap equation is

$$M(p^2) = \frac{3\alpha}{2\pi^2}\int_0^{\Lambda^2} dk^2 k^2 \frac{M(k^2)}{k^2 + M^2(k^2)}\int_0^\pi d\psi \frac{\sin^2\psi}{q^2}, \quad (4)$$

where Λ is an ultraviolet cut-off. DMG in QED arises when the coupling constant reaches a critical value ($\alpha_c = \pi/3$) as can be seen by solving numerically Eq.(2). In a vicinity of the

Journal of Physics: Conference Series **468** (2013) 012010 doi:10.1088/1742-6596/468/1/012010

critical coupling α_c, the generated mass $m_d = M(0)$ obeys the Miransky scaling law [3]

$$\frac{\Lambda}{m_d} = \exp\left(\frac{A}{\sqrt{\frac{\alpha}{\alpha_c} - 1}} - B\right). \tag{5}$$

In order to corroborate this scaling relation, we solve numerically the Eq.(2) for several values of the coupling. Results are shown in Fig. 1. The curve for QED agrees with the Miransky law with $A \simeq \pi$ and $B \simeq 2$.

Another important non-perturbative aspect to analyze is confinement. It is said that a theory is confined if we need an infinite amount of energy to take apart a two-particle bound state. To perform our confinement test, we invoke the violation of reflection positivity in the following way: If the vector part of the propagator

$$\sigma_v(p^2) = \frac{1}{p^2 + M^2(p^2)}$$

presents a inflection point, then we have confinement; the inflection point guarantees that the Fourier transform of σ_v flips sign in coordinate space, hence breeching reflection positivity. In Fig. 2 we plot the derivate of the σ_v. We can note from the absence of local extrema that we do not have confinement as expected for a theory with a static electromagnetic potential which behaves like $1/r$.

4. Gap equations in QED3 & Graphene

4.1. QED3

For QED3, the gap equation is

$$M(p) = \frac{\alpha}{\pi^2} \int \frac{d^3k}{q^2} \frac{M(k)}{(k^2 + M^2(k))}. \tag{6}$$

By means of a numerical analysis of Eq.(4), we can verify that there is not critical coupling for DMG to take place, opposedly to QED. This means that if have DMG for one value of α, we have it for all values of the coupling. Moreover, we observe that the mass function scales like

$$\alpha M(p; \alpha) = M(p/\alpha; 1) , \tag{7}$$

such that the scaling of the dynamically generated mass is lineal with the coupling, as can be seen in Fig. 1.

On the other hand, performing the same confinement test as in QED, we notice that this theory is confined as we expected for a planar theory with a logarithmic potential (Fig. 3).

4.2. Graphene

The low energy effective theory of graphene corresponds to a peculiar form of QED in which the charge carriers, the graphinos, are restricted to move in a plane, while the photons are freely moving in space. The question now is: What differences and similarities can we find between these theories and the dynamical generation of masses and confinement in graphene? This question is closely related to the semi-metal–insulator transition in graphene. To answer this, we start from the gap equation proposed by Gusynin *et al.* [4]

$$M(p) = \frac{2\lambda}{\pi} \int_0^{\Lambda} \frac{kM(k)\, dk}{\sqrt{k^2 + M^2(k)}} \mathcal{K}(p,k), \qquad \lambda = \frac{\alpha}{2(1 + \pi\alpha/2)}, \tag{8}$$

where the lineal kernel is

$$\mathcal{K} = \frac{\pi}{2}\left(\frac{\theta(p-k)}{p} + \frac{\theta(k-p)}{k}\right).$$

This equation takes into account the one-loop corrections to the Coulomb interaction and can be solved by similar techniques. A numerical analysis shows that, just as in QED, there is a critical coupling $\lambda_c \sim 1/4$ above which mases are dynamically generated. Interestingly, the generated mass follows a Miransky-type scaling law. This means that the scenario of mass generation in graphene is similar to that of QED.

On the other hand, we perform the confinement on the solution to the gap equation and observe confinement just, like in QED3 [3].

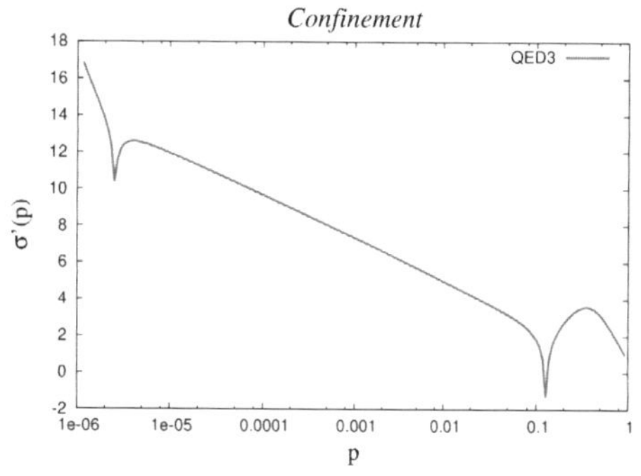

Figure 3. Confinement Test QED3.

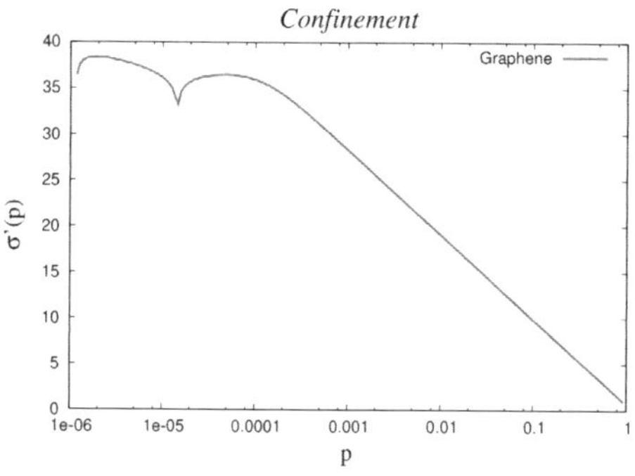

Figure 4. Confinement Test Graphene.

5. Conclusions

In this work, we have explored similarities and difference of the gap equation in QED, QED3 and the low-energy effective theory of graphene regarding how dynamical mass generation and confinement are linked. We have found that for graphene, the scenario for confinement is similar to QED3. However, considering that photons are not restricted to the plane, the theory requires a critical value for coupling to generate masses, just like in QED. Thanks to this critical value a gap is opened and the graphene goes from a semimetal to a insulator.

Acknowledgments

We acknowledge to CONACyT and CIC. SH thanks the organizing committee for support.

References

[1] Novoselov K S et al 2005 Nature 438 197;
Zhang Y et al 2005 Nature 438 201;
Gusynin V P and Sharapov S G 2005 Phys. Rev. Lett. 95 146801.
[2] Faisal F H M Adiabatic solution of Dirac equation of graphinos in an intense electromagnetic field and emission of high order harmonics near the Dirac points arXiv:1102.1677 [cond.mat.mes-hall];
Hernández-Ortíz S, Murguía G and Raya A 2012 J. Phys. Condens. Matter 24 015304.
[3] Hernández-Ortíz S 2012, Master Thesis IFM-UMSNH, in progress.
[4] Gamayun OV, Gorbar EV, Gusynin VP 2010, Phys. Rev. B81, 075429.

Journal of Physics: Conference Series **468** (2013) 012011 doi:10.1088/1742-6596/468/1/012011

Zero field Quantum Hall Effect in QED3

K Raya[1], S Sánchez-Madrigal[1,2], A Raya[1]

[1]Instituto de Física y Matemáticas, Universidad Michoacana de San Nicolás de Hidalgo, Edificio C-3, Ciudad Universitaria, Morelia, Michoacán 58040, Mexico
[2]Instituto de Física, Univesidad Nacional Autónoma de México (UNAM)
Circuito de la Investigación Científica, Ciudad Universitaria, Mexico D.F. 04510, Mexico.

E-mail: khepani@ifm.umich.mx, madrigals@fisica.unam.mx, raya@ifm.umich.mx

Abstract. We study analytic structure of the fermion propagator in the Quantum Electrodynamics in 2+1 dimensions (QED3) in the Landau gauge, both in perturbation theory and nonperturbatively, by solving the corresponding Schwinger-Dyson equation in rainbow approximation. In the chiral limit, we found many nodal solutions, which could be interpreted as vacuum excitations. Armed with these solutions, we use the Kubo formula and calculate the filling factor for the zero field Quantum Hall Effect.

1. Introduction

One of the most remarkable phenomena in condensed matter physics is the Quantum Hall Effect (QHE), both integer [1] and fractionary [2]. Such phenomenon can be described entirely in terms of a relativistic theory such as quantum electrodynamics [3]. The transverse conductivity for relativistic fermions restricted to move in a plane at zero magnetic field becomes

$$\sigma_{xy} = -\nu \frac{e^2}{2\pi} , \tag{1}$$

in units where $c = \hbar = 1$. Here, e^2 is the electron charge and the number ν is the filling factor, which can be computed from the Kubo formula [4]

$$\nu = \frac{1}{6\pi^2}\epsilon^{\mu\nu\lambda} \int d^3p \, \mathrm{Tr}\left[S(p)\left(\partial_\mu S^{-1}(p)\right) S(p)\left(\partial_\nu S^{-1}(p)\right) S(p)\left(\partial_\lambda S^{-1}(p)\right)\right] , \tag{2}$$

where $S(p)$ is the electron propagator and $\partial_\mu = \partial/\partial p^\mu$. The fermion propagator is expressed in its most general form as

$$S(p) = \frac{F(p)}{\not{p} - M(p)} , \tag{3}$$

where $F(p)$ and $M(p)$ are the renormalization and mass functions respectively. It was shown by Acharya-Narayana [4] and Jellal [5] that, once wavefunction renormalization effects are neglected ($F(p) = 1$), the general expression for the filling factor acquires the form

$$\nu = \frac{1}{6\pi^2} \int d^3p \, \frac{3M(p) + 2p^2 M'(p)}{[p^2 + M^2(p)]^2} . \tag{4}$$

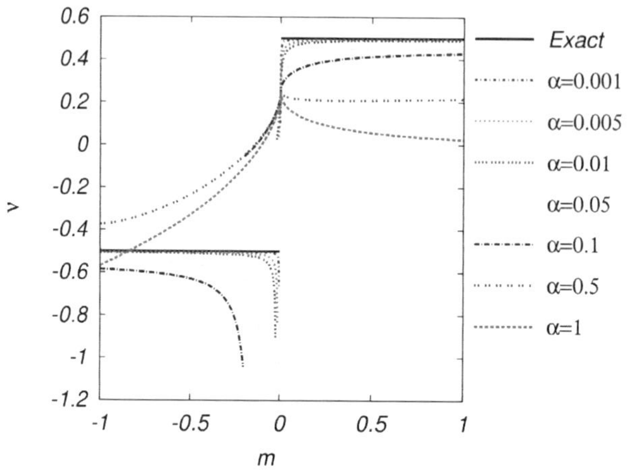

Figure 1. Filling factor ν at 1-loop, as a function of m for different values of the coupling α.

Our purpose is to check the validity of (4) using perturbation theory considering the free propagator and its 1-loop correction. Non-perturbatively, we solve the Schwinger-Dyson Equation (SDE) for the electron propagator in the rainbow truncation. Multiple nodal solutions are found. We calculate the filling factor ν associated to these solutions and interpret them as describing vacuum excitations.

2. Free Fermion Propagator

The inverse free fermion propagator is $S_0^{-1}(p) = \not{p} - m$, where m is the bare mass and we select the Dirac matrices in the form

$$\gamma^0 = \sigma^3 \ , \ \gamma^k = i\sigma^k \ , \tag{5}$$

with $k = 1, 2$ and σ_j are the Pauli matrices. The mass term violates parity [3], and therefore produces a non-trivial filling factor

$$\nu = \frac{1}{2}\frac{m}{|m|}. \tag{6}$$

This value corresponds to the filling factor of the zero field QHE [3]. In the massless limit, it is said that at zero magnetic field, vacuum is half-filled.

3. Fermion Propagator in Perturbation Theory

At 1-loop order, the fermion propagator in covariant gauges is

$$\begin{aligned} \frac{1}{F_1(p)} &= 1 - \frac{\alpha\xi}{2\pi^2 p^2}\int d^3k\ \frac{(k^2+p^2)(k\cdot p) - 2k^2p^2}{q^4(k^2+m^2)} \ , \\ \frac{M_1(p)}{F_1(p)} &= m - \frac{\alpha(2+\xi)m}{2\pi^2}\int d^3k\ \frac{1}{q^2(k^2+m^2)} \ , \end{aligned} \tag{7}$$

where $\alpha = e^2/(4\pi)$, as usual and ξ is the covariant gauge parameter. Performing the integrations involved, we obtain [6]

$$\begin{aligned} \frac{1}{F_1(p)} &= 1 + \frac{\alpha\xi}{2p^2}[m - (m^2 - p^2)I(p;m)] \ , \\ \frac{M_1(p)}{F_1(p)} &= m[1 + \alpha(\xi+2)I(p;m)] \ , \end{aligned} \tag{8}$$

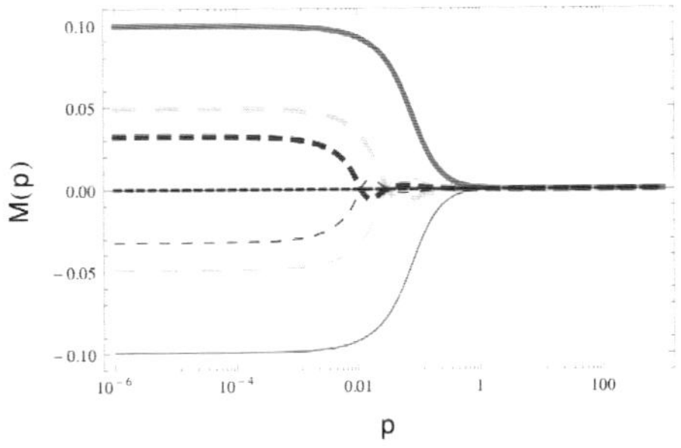

Figure 2. Multiple nodal solutions to the gap equation (11).

where $I(p;m) = (1/p)\arctan(p/m)$. In the Landau gauge ($\xi = 0$), again $F_1(p) = 1$, thus we can use the Kubo formula (4).

Inserting the mass function $M_1(p)$, we obtain the filling factor as function of the electron mass for various values of the coupling α, as shown in Fig. 1. It is observed that when $\alpha \ll 1$, there are no significant deviation from the half-filling value. Deviations are more evident when α is larger, but in this regime, the perturbative behavior is compromised.

4. Dynamical Mass Generation and the Filling Factor

In this section we study the non-perturbative structure of the fermion propagator through the corresponding SDE

$$S^{-1}(p) = S_0^{-1}(p) - 4\pi\alpha \int \frac{d^3k}{(2\pi)^3} \Gamma^\mu(k;p) S(k) \gamma^\nu \Delta_{\mu\nu}(q) , \tag{9}$$

where $q = k - p$, $\Gamma^\mu(k;p)$ is the fermion-photon vertex, which from now onward we consider bare, $\Gamma^\mu(k;p) = \gamma^\mu$, and the full photon propagator is given by

$$\Delta_{\mu\nu}(q) = -\left[\frac{G(q)}{q^2}\left(g_{\mu\nu} - \frac{q_\mu q_\nu}{q^2}\right) - \xi\frac{q_\mu q_\nu}{q^4}\right] , \tag{10}$$

where $G(q) = 1/(1+\Pi(q))$ is the wavefunction renormalization function of the photon, with $\Pi(q)$ is the vacuum polarization scalar. $G(q) = 1$ corresponds to the bare photon propagator, which is our choice for the truncation of SDE. Dynamical mass generation with reducible fermions has been studied in [7]. It is assumed that along with chiral symmetry, parity is also dynamically broken. This view, however, has been challenged [8]. We stick to the interpretation of [7]. Working in Landau gauge, Eq. (9) reduces to

$$M(p) = 8\pi\alpha \int \frac{d^3k}{(2\pi)^3} \frac{M(k)}{q^2(k^2 + M^2(k))} , \tag{11}$$

in Euclidean space. We have found multiple nodal solutions to the above equation in the chiral limit [9], as shown in Fig. 2. These solutions posses a mirror image around the horizontal axis. Such solutions gradually disappear when the bare mass m is increased and it reaches a critical mass (m_{crit}). Above m_{crit}, only the non-nodal solution remains [9].

We insert these solutions into equation (4) and we obtain the filling factor ν as a function of the number of nodes. Results are displayed in Fig. 3. For the non-nodal solution, the filling factor is slightly smaller that $1/2$. The difference arises from the assumptions behind the rainbow

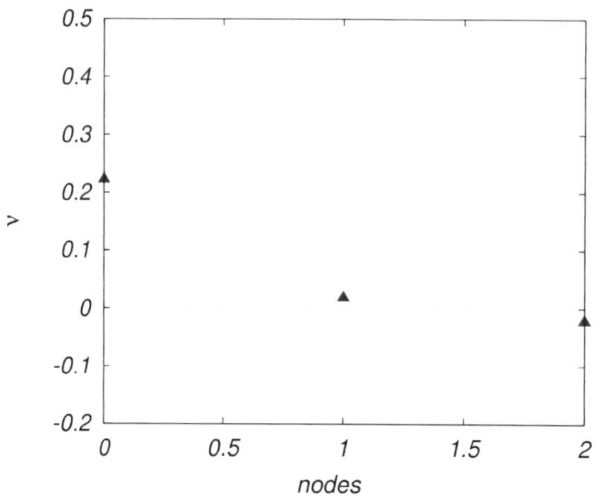

Figure 3. Filling factor ν for the many nodes solutions.

approximation. For higher nodes, we observe that the filling factor, in absolute value, diminishes. This allows to interpret the nodal solutions as vacuum excitations; there are less of those states populating the vacuum.

5. Conclusions

In this study, we arrived to the following conclusions:

- In perturbation theory, there are no significant corrections to the Kubo formula so long as the coupling remains small.
- Noded solutions show that there is a reduction in ν as the number of nodes increases. Therefore, these solutions could be interpreted as vacuum excitations.

All these things are being checked in detail [10].

Acknowledgments

We acknowledge support from CONACyT and UMSNH. KR acknowledges the organizing committee for financial support to attend this event.

References

[1] von Kiltzing K, Dorda G and Pepper M 1980 Phys. Rev. Lett. 45 494.
[2] Tsui D C, Stormer H L and Gossard A C 1982 Phys. Rev. Lett. 48 1559.
[3] Raya A and Reyes E 2008, Jour. Phys. A 41, 355401.
[4] Acharya R and Narayana P 1992, Nuovo Cim. B107 351. (1990).
[5] Jellal A 1998. Int. Jour. Theo. Phys. 37, 2187.
[6] Bashir A,2008 Raya A and Sánchez S , J. Phys. A. 41, 505401.
[7] Hoshino Y and Matsuyama T 1989, Phys. Lett. B222, 493.
[8] Lo P M and Swanson E S 2011, Phys. Rev. D83 065003.
[9] Raya K 2010, Undergraduate Thesis. UMSNH, Mexico.
[10] Raya A, Raya K and Sánchez S 2012, *Work in progress*.

Journal of Physics: Conference Series **468** (2013) 012012 doi:10.1088/1742-6596/468/1/012012

A calculation for $Br(Z' \to tth)$ in a B-L model

J M Hernández López[1] **and J A Orduz-Ducuara**[2]

Facultad de Ciencias Físico Matematicas, Benemérita Universidad Autónoma de Puebla, 18 Sur y Av. San Claudio, C.P. 72570 Puebla, Pue., Mexico

E-mail: [1]`javierh@fcfm.buap.mx`, [2]`jaorduz@alumnos.fcfm.buap.mx`

Abstract. Models with extra gauge bosons often requires an extended Higgs sector, with properties that deviate from the Standard Model (SM) case. These modifications would affect the strength of the signals that are used to search for the Higgs spectrum at present and future colliders. However, it is also possible that new signals could arise in these models. In particular, the existence of such new gauge bosons could also be used to provide new production mechanisms. In this paper we will study a useful decay for a new gauge boson, Z', with SM-like couplings, namely $Z' \to f\bar{f}h$, through its branching ratio. We conceive this process as a higgsses possible source, steaming from the fact that the LHC is going to be a natural source of new gauge neutral bosons, in the case nature agrees with us about its existence. We found a corresponding branching ratio of order 10^{-3}, under certain scenaries. In particular we have used $m_h = 125\,\text{GeV}$ as the last results.

1. Introduction

After so many years of preparation, the LHC started to test in very significant ways the Higgs sector of the SM. We know now that the SM Higgs boson has $m_h = 125$ GeV, which agrees quite well with the range preferred by the analysis of electroweak precision tests [1, 2]. The LHC has also provided important bounds on the scale of new physics beyond the SM.

One of the simplest extensions of the SM, with broad motivations, is the presence of new gauge bosons, which could arise within a variety of contexts, ranging from simple extra U(1) gauge symmetries [3], left-right models [3–6], GUTs, [7], and even string theory [8–10]. In particular, the extension of the SM with an extra $U(1)'$ has been studied extensively. Predictions for detection at LHC as well as other phenomenological studies have been presented [3–6, 8–10].

Here, we would like to know what would be the consequences for the Higgs sector once such extra neutral gauge boson would be detected. In the first place, we know that such models with extra gauge bosons, often require an extended Higgs sector, with properties that desviate from the Standard Model case. These modifications will change the strength of Higgs interactions with SM particles, which in turn would induce modifications of the standard signals that are used to search for the Higgs spectrum at present and future colliders. However, the new gauge boson could also induce new signals for the Higgs bosons.

In this work, we calculate the branching ratio for the 3-body decay of the Z' into a Higgs boson and a top anti-top pair, i.e. $Z' \to t\bar{t}h$. We find that it is possible to get $\mathbf{Br}'s$ (Branching Ratios) as high as 10^{-3}, which could be feasible to be studied at LHC.

Journal of Physics: Conference Series **468** (2013) 012012 doi:10.1088/1742-6596/468/1/012012

2. A simple model with an extra Z'

In this section we sumarize the main features of the model [11–16].

$$\mathcal{L} = (D^{\mu}\Phi)^{\dagger}(D_{\mu}\Phi) + (D^{\mu}\chi)^{\dagger}(D_{\mu}\chi) - V(H,\chi)$$

the covariant derivate is [12],

$$D_{\mu} = \partial_{\mu} + i\left[g\frac{\sigma^a}{2}W^a_{\mu} + Q^Y_i B_{\mu} + g_P Y^P_i B'_{\mu}\right]$$

for the singlet under SU(2), we obtain:$(D^{\mu}\chi)^{\dagger}(D_{\mu}\chi) = \frac{1}{2}\partial^{\mu}h'\partial_{\mu}h' + \frac{1}{2}(h'+x)^2(2g'_1 B'\mu)^2$ and the potential is [12]: $V(H,\chi) = m^2 H^{\dagger}H + \mu^2|\chi|^2 + \lambda_1(H^{\dagger}H)^2 + \lambda_2|\chi|^4 + \lambda_3 H^{\dagger}H|\chi|^2$

There are several bounds for the Z' mass and its couplings, as the mixing angle, for different scenarious and processes [13, 17]. Some **Br** for different channels have been studied in [18], they use $M_{Z'} = 1\,\text{TeV}$. By the other side, Ref. [19] exclude $M_{Z'} < 1140\,\text{GeV}$ with standard-model-like couplings and $M_{Z'} < 887\,\text{GeV}$ in superstring-inspired models. Hencefor if a new gauge boson exists, its mass would have $M_{Z'} \sim \mathcal{O}\,(\text{TeV})$ because it is necessary to avoid fine-tunnings.

The mixing angle (θ'), between Z and Z' states, has a definite bound from precision measurement observables, namely, $|\theta| \lesssim 10^{-3}$ [20]. The coupling ${g'}_1$ has values model-dependent but it is bounded by $\mathcal{O}\left(10^{-1}, 10^{-2}, 10^{-3}\right)$; for instance, see [20]. In [11], they found a dinamical way to constrain ${g'}_1$ for the $B-L$ model. The general relation for the coupling and the mass is given by, $M_{Z'} > {g'}_1 6\,\text{TeV}$ It is possible to find a light new gauge boson if its couplings to the SM field are small [14]. We refer to the reader to [15] for a deeper insight.

3. $Z' \to f\bar{f}h$ Branching ratio

We calculate the $\mathbf{Br}(Z' \to f\bar{f}h)$ with the following Feynman diagrams,

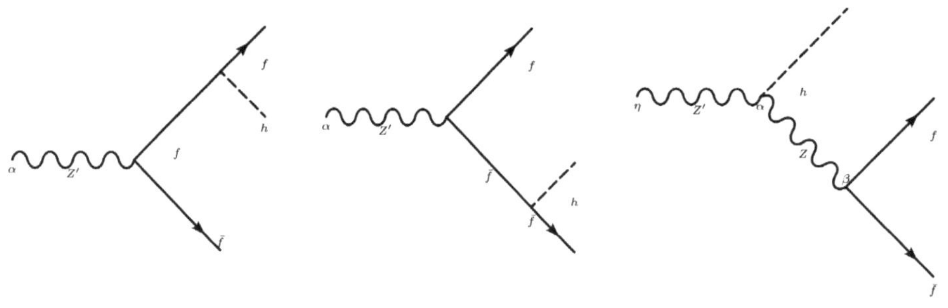

Figure 1. Feynman diagram for the process

taking into account all the interference terms in the calculations, we get,

$$\begin{aligned}\frac{d\Gamma(Z'\to f\bar{f}h)}{dx_f dx_{\bar{f}}} = {} & \frac{M_{Z'}}{256\pi^3}\left[\frac{g^4 M^2_{Z'}}{2\cos^4\theta'\Gamma^2_f}\left[\left(\mu_f + 2\mu_h + x_{\bar{f}} - 1\right)\left(\frac{1}{2}\left({g^f_A}^2 + {g^f_V}^2\right)\left(-2\mu_f + \mu_h + x_f + x_{\bar{f}} - 1\right) - 4\mu_f {g^f_V}^2\right) + \right.\right. \\ & \left. \frac{\mu_f}{2}\left(5{g^f_V}^2 - 3{g^f_A}^2\right)\left(-2\mu_f + \mu_h + x_f + x_{\bar{f}} - 1\right) + 2\mu_f^2\left({g^f_A}^2 - {g^f_V}^2\right)\right] + \\ & \frac{g^4 M^2_{Z'}}{2\cos^4\theta'\Gamma^2_{\bar{f}}}\left[\frac{1}{2}\left({g^f_A}^2 + {g^f_V}^2\right)(-2\mu_f + \mu_h + x_f + x_{\bar{f}} - 1)(4\mu_f + 2\mu_h + x_f - 1) - \right. \\ & \left. 2\mu_f\left({g^f_A}^2 - {g^f_V}^2\right)(4\mu_f + 3(2\mu_h + x_f - 1))\right] +\end{aligned}$$

$$\begin{aligned}
&\frac{8g^2v^2A_1^2M_{Z'}^2}{\cos^2\theta M_Z^2\Gamma_Z^2}\left[2\mu_f\left({g_V^f}^2-{g_A^f}^2\right)+\frac{1}{2}\left({g_A^f}^2+{g_V^f}^2\right)(-2\mu_f+\mu_h+x_f+x_{\bar f}-1)\right]+\\
&\left(\frac{8g^2M_{Z'}}{\cos^2\theta'}\right)^2\left(\frac{1}{\Gamma_f\Gamma_{\bar f}}\right)\times\\
&\Big[2\mu_f\left(\frac{1}{2}\left({g_V^f}^2-3{g_A^f}^2\right)\left(-2\mu_f+\mu_h+x_f+x_{\bar f}-1\right)+\left({g_A^f}^2-{g_V^f}^2\right)(2\mu_f+2\mu_h+x_f-1)\right)+\\
&\left({g_A^f}^2+{g_V^f}^2\right)(2\mu_f+2\mu_h+x_f-1)\left(\mu_f+2\mu_h+x_{\bar f}-1\right)\Big]+\\
&\left(\frac{g^3vm_fA_1M_{Z'}}{\cos^3\theta'}\right)\left(\frac{1}{\Gamma_fM_Z\Gamma_Z}\right)\Big[-\left({g_A^f}^2+{g_V^f}^2\right)\left(\mu_f+2\mu_h+x_{\bar f}-1\right)-\frac{1}{2}\left({g_V^f}^2-3{g_A^f}^2\right)\times\\
&(-2\mu_f+\mu_h+x_f+x_{\bar f}-1)-2\mu_f\left({g_A^f}^2-{g_V^f}^2\right)\Big]+\left(\frac{4g^3m_fvA_1M_{Z'}}{\cos^3\theta'}\right)\times\\
&\left(\frac{-\left({g_A^f}^2+{g_V^f}^2\right)\left(\mu_f+2\mu_h+x_{\bar f}-1\right)-\frac{1}{2}\left({g_V^f}^2-3{g_A^f}^2\right)(-2\mu_f+\mu_h+x_f+x_{\bar f}-1)-2\mu_f\left({g_A^f}^2-{g_V^f}^2\right)}{M_Z\Gamma_Z\Gamma_f}\right)\Big]
\end{aligned} \quad (1)$$

where the integration limits, for the scaled energy variables, $x_i = 2E_i/M_{Z'}$, are:

$$2\sqrt{\mu_f} \leq \quad x_{\bar f} \quad \leq 1-\mu_h-2\mu_f$$
$$\frac{1}{2}\frac{A-B}{1-x_{\bar f}+\mu_f} \leq \quad x_f \quad \leq \frac{1}{2}\frac{A+B}{1-x_{\bar f}+\mu_f}$$

and, $A=(2-x_{\bar f})(1+2\mu_f-\mu_h-x_{\bar f})$, $B^2=(x_{\bar f}^2-4\mu_f)\left((1-x_{\bar f}-\mu_h)^2-4\mu_f^2\right)$, $\mu_i=\frac{m_i^2}{M_{Z'}^2}$, $(i = f,\bar f)$ and $\sum_{i=1}^3 x_i = 2$.

A_1 is a coefficient comes from the vertex hZZ'; where $A_1 = 2\left(\frac{\cos\alpha}{4}\Theta_{hZZ'} - r\sin\alpha{\Theta'}_{hZZ'}\right)$, $\Theta_{hZZ'} = -\sin 2\theta'(g^2+g'^2)$, ${\Theta'}_{hZZ'} = 4{g'}_1\sin 2\theta'$ and $r=\frac{v}{v'}$ for a pure $B-L$ model.

We must stress that even when the $Z-Z'$ mixing angle it is often set to zero, there are plausible models that allow a 10^{-3} level, value which we use (see [8–10]). In order to be confident about the certainty of a zero mixing, we studied the cases for $10^{-4,-5}$ for the mixing angle. We used the neutrino's masses from [16].

Recent results have set a strict bound for the Higgs boson mass, at 125 GeV [1, 2, 21]. This is the nominal value we use.

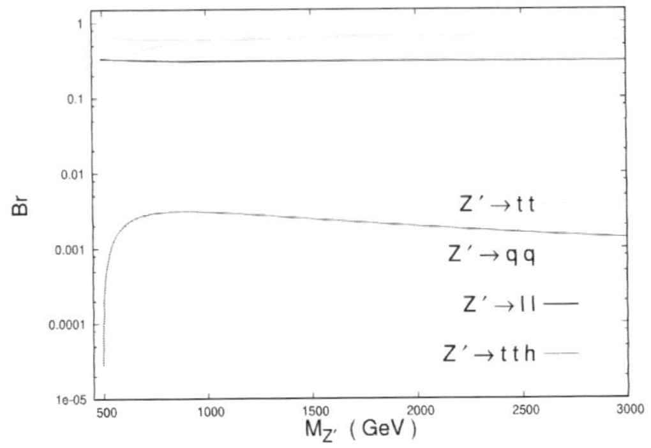

Figure 2. We have set $v' = 1\,\mathrm{TeV}$ and $\theta' = 1\times10^{-4}, \alpha = \frac{\pi}{2}$

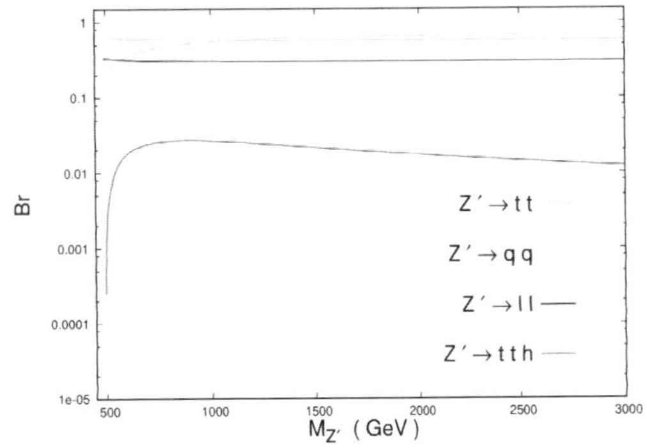

Figure 3. We have set $v' = 3\,\mathrm{TeV}$ and $\theta' = 1\times10^{-5}$. $\alpha = \frac{\pi}{2}$.

In figs. (2) and (3), we show the Branching ratio (**Br**) for different channels as a function of the new gauge boson mass. We have considered $q = u, c, d, s, b.$ and $l = e, \mu, \kappa, \nu_e, \nu_\mu, \nu_\kappa$. We use for definitiveness SM-like couplings for Z'-SM fermions (a $B-L$ like scenarious). We explored the **Br** for $\theta' = 1\times10^{-4}, 1\times10^{-5}$ and 1×10^{-6} with $v' = 1\,\mathrm{TeV}, 2\,\mathrm{TeV}$ and $3\,\mathrm{TeV}$; however there were no significant differences.

4. Conclusions

A **Br** of 10^{-3} is well within the reach of the LHC studies. So, in the case of a Z' discovery, this process would be a good candidate to search and study a new gauge boson, of course in the case the nature choices as the next step a sequential model. Otherwise, in other kind of models, it could be a good source of Higgses, in order to clarify if the boson just announced it is effectively the SM boson or not.

CMS or ATLAS [1, 2] could be able to detect a new gauge boson in the TeV range. As well as to measure and reconstruct both a pair of top and a pair of photons, as could be the Z' signature in our case. The QCD background for the final states $\bar{t}t\gamma\gamma$ has a continuum spectrum for the photons, while the Higgs signal we are looking for has been refined meanwhile. In this way, the possibility of the Z' reconstructing is a reachable level from the invariant mass of the final states. A more detailed and exhaustive work is in order, and it will be published elsewhere.

Acknowledgments

We acknowledge support from CONACYT-SNI (Mexico).

References

[1] Chatrchyan S *et al.* (CMS Collaboration) 2012 *Phys.Lett.B* (*Preprint* 1207.7235)
[2] Aad G *et al.* (ATLAS Collaboration) 2012 (*Preprint* 1207.7214)
[3] Pati J C and Salam A 1974 *Phys.Rev.* **D10** 275–289
[4] Mohapatra R N and Pati J C 1975 *Phys.Rev.* **D11** 566–571
[5] Mohapatra R N and Senjanovic G 1981 *Phys.Rev.* **D23** 165
[6] Mohapatra R N and Senjanovic G 1980 *Phys.Rev.Lett.* **44** 912
[7] Kane Gordon L e 1998
[8] Cvetic M and Langacker P 1996 *Phys.Rev.* **D54** 3570–3579 (*Preprint* hep-ph/9511378)
[9] Cvetic M and Langacker P 1996 *Mod.Phys.Lett.* **A11** 1247–1262 (*Preprint* hep-ph/9602424)
[10] Cleaver G, Cvetic M, Espinosa J, Everett L, Langacker P *et al.* 1999 *Phys.Rev.* **D59** 115003 (*Preprint* hep-ph/9811355)
[11] Basso L, Moretti S and Pruna G M 2012 *J.Phys.G* **G39** 025004 (*Preprint* 1009.4164)
[12] Basso L 2006-2007 A minimal extension of the standard model with "b-l" gauge symmetry
[13] Sahoo S and Maharana L 2004 *Pramana* **63** 491–507
[14] Emam W and Khalil S 2007 *[hep-ph/0704.1395] Eur. Phys. J.* **C55** 625–633 (*Preprint* [hep-ph/0704.1395])
[15] Langacker P 2009 *[hep-ph/0801.1345] Rev. Mod. Phys.* **81** 1199–1298 (*Preprint* [hep-ph/0801.1345])
[16] Langacker P 2009 (*Preprint* [hep-ph/0911.4294])
[17] Abbiendi G *et al.* (OPAL Collaboration) 2004 *Eur.Phys.J.* **C33** 173–212 (*Preprint* hep-ex/0309053)
[18] Barger V, Keung W Y and Yu C T 2011 *Phys.Lett.* **B698** 243–250 (*Preprint* 1102.0279)
[19] Chatrchyan S *et al.* (CMS Collaboration) 2011 *JHEP* **1105** 093 (*Preprint* 1103.0981)
[20] C Balazs J L Diaz-Cruz H J H T M P T and Yuan C P
[21] Erler J 2012 (*Preprint* [hep-ph/1201.0695])

Journal of Physics: Conference Series **468** (2013) 012013 doi:10.1088/1742-6596/468/1/012013

The low-energy extensions of the Pierre Auger Observatory

E Varela[1] for the Pierre Auger Collaboration[2]

[1]Facultad de Ciencias Físico-Matemáticas, BUAP, Av. San Claudio y 18 Sur, Col. San Manuel Puebla, México

[2] Observatorio Pierre Auger, Av. San Martín Norte 304, 5613 Malargüe, Argentina (Full author list: *http : //www.auger.org/archive/authors_2012_09.html*)

E-mail: evarela21282@hotmail.com

Abstract. The Pierre Auger Observatory was designed to measure ultra-high energy cosmic rays above 10^{18}eV with high accuracy using a hybrid air shower detection technique. A Surface Detector (SD) with 1600 water-Cherenkov stations on a 1500 m triangular grid covers an area of 3000 km^2. The atmosphere above the array is viewed by a Fluorescence Detector (FD) with 24 telescopes at 4 sites in the periphery of the SD. As an enhancement to this baseline design, in order to reach a lower energy threshold below 10^{17}eV, the Collaboration has implemented extensions to the Observatory. The SD extension is AMIGA (Auger Muons and Infill for the Ground Array), an infilled area with detectors at a smaller spacing than in the main array and with buried scintillator counters. The FD is complemented by HEAT (High Elevation Auger Telescopes), with 3 additional telescopes that are tilted upwards to extend the elevation range. The importance of these extensions of the Auger Observatory is that they allow to study the energy range where the transition from a Galactic to an extra-Galactic origin of cosmic rays may occur. The current status and initial results from these extensions are discussed.

1. Introduction

The origin and nature of the Ultra-high Energy Cosmic Rays (UHECR) has been a mystery since they were discovered in 1962 [1]. These cosmic rays reach energies above 10^{20} eV [2], questions about how and where they are accelerated, have been asked for decades. Our knowledge on Ultra High Energy Cosmic Rays is limited, one reason is the very low flux of UHECRs at Earth (see figure 1). Large detectors are needed to collect information of the Extensive Air Showers (EAS) produced by the primary cosmic ray particles in order to infer properties of the primary particle, such as its energy, its arrival direction, and its composition, with enough statistics. The flux of cosmic rays decreases with energy and can be described by a power law like $\frac{dN}{dE} \sim E^{-\alpha}$ with spectral index α between 2.6 and 3.2. One can find some small but relevant breaks at the energy spectrum, the knee at $\sim 4 \times 10^{15}$ eV, where α changes from 2.7 to 3.1, the second knee (slightly above 10^{17} eV) and the ankle at 4×10^{18} eV (see figure 1) [3] . The transition from galactic to extragalactic cosmic rays may happen between the second knee and ankle, therefore, cosmic rays in this energy range are most likely generated by different astrophysical sources, and their composition may change as well.

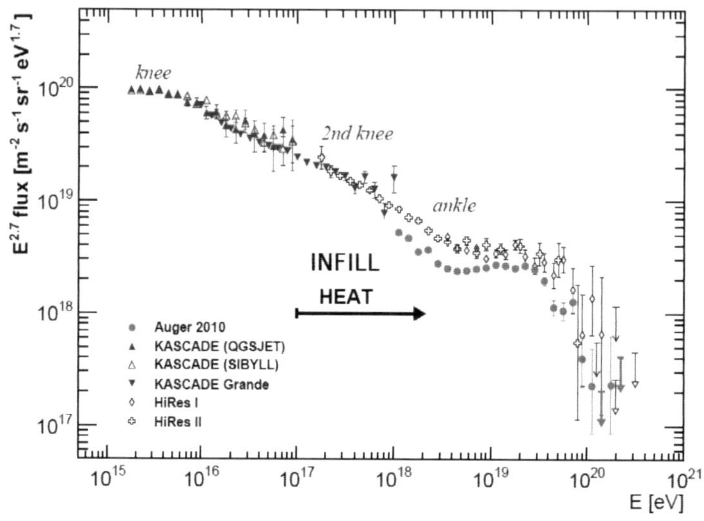

Figure 1. Energy spectrum of cosmic rays. The particle flux is scaled with $E^{-2.7}$ to evidence the spectral features (knee, second knee and ankle). The arrow starting from 10^{17} eV indicates the energy range where the Auger Observatory extends its studies of cosmic rays through HEAT and AMIGA. The figure was modified and taken from [4]

2. The Pierre Auger Observatory

The Pierre Auger Observatory [5], is located near the city of Malargüe, Argentina, at latitude 35.2° South and an altitude of 1400 m.a.s.l.. Its baseline design, inaugurated in November 2008, was completed in May 2008. The Observatory has been taking data since January 2004 during its construction phase. The Pierre Auger Observatory was designed to measure properties of cosmic rays with energy above 10^{18} eV such as their flux, arrival directions and mass composition with unprecedented statistics. The Observatory is a hybrid cosmic ray detector, that consists of two complementary detectors that can observe, in coincidence, the longitudinal development of the shower of particles with air-fluorescence telescopes and its pattern when it reaches the Earth surface, where they can be spread over several kilometres. The Surface Detector [6] determines the particle densities at ground. It is composed of 1600 Cherenkov stations separated by 1.5 km, covering a total area of 3000 km^2. Surface detectors use generally scintillators or water-Cherenkov stations as their detector stations. Water Cherenkov stations have broader angular coverage, compared to thin scintillators which are usually more difficult to build and deploy. Both methods are robust and reliable, well known, detectors adequate to instrument the required large areas. The Fluorescence Detector [7] consists of 4 buildings on the outskirts of the SD array, each one with 6 telescopes that overlook the array covering 30^0 in elevation and 180^0 in azimuth. It collects the fluorescence light emitted by atmospheric nitrogen molecules excited as the shower is crossing the atmosphere. The FD measures the longitudinal profile of the shower. Fluorescence detectors record the near UV light emitted by deexcitation of N_2 molecules previously excited by the shower particles. They provide a direct measurement of the longitudinal shower development. This is crucial, because the integral of the profile is a direct measurement of the energy deposited in the atmosphere. Moreover, the depth of the shower maximum is also a good indicator of the primary mass. Three factors limit the ability to study cosmic ray primary composition: intrinsic fluctuations on shower development, finite detector resolution and systematics and limited knowledge of high energy hadronic interactions. The layout of the Southern Observatory is shown in figure 2

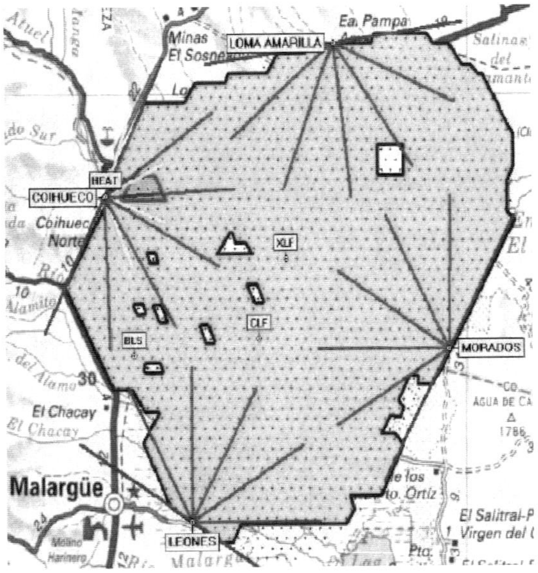

Figure 2. Layout of the Pierre Auger Observatory. The points are the water-Cherenkov stations of the Surface Detector and the green rays indicate the field of view of each of the 24 telescopes of the Fluorescence detector. The 4 sites for the fluorescence telescope are indicated by name.

2.1. Surface Detector

The Surface Detector measures the front of the shower as it reaches ground. The detectors activated by the event record the particle density and the time of arrival.

The cell unit (see figure 3) of the Surface Detector of the Pierre Auger Observatory is a water Cherenkov counter. Each counter is a polyethylene station of cylindrical shape with size 10 $m^2 \times 1.2$ m filled with purified water. Cherenkov light produced by charged particles of the showers is detected by three 9" photomultipliers. Each unit is autonomous with a battery and a solar panel.

Stations are calibrated using the signals of atmospheric muons, a well known uniform background across the array. The signal of a muon is proportional to the geometric path length. A test station was used to establish the relation between the signal of down-going vertical and central muons (VEM) and the peak of the histogram obtained from omni directional muons crossing the station. Each station is calibrated matching the photomultipliers gain to obtain the expected trigger rate over a given VEM threshold. This procedure allows to calibrate the stations with respect to the absolute value of the VEM with an overall precision of 2% [8].

2.2. Fluorescence Detector

The fluorescence detector consists of 24 telescopes in four sites, located on small elevations on the perimeter of the array (figure 2). The telescopes measure the shower longitudinal development in the atmosphere by collecting the fluorescence light emitted by the atmospheric nitrogen molecules excited by the charged particles. Each telescope is composed by an aperture system, a spherical mirror, and a 440 photomultiplier camera at the focal plane. A schematic view of the arrangement can be seen in figure 4.

Journal of Physics: Conference Series **468** (2013) 012013 doi:10.1088/1742-6596/468/1/012013

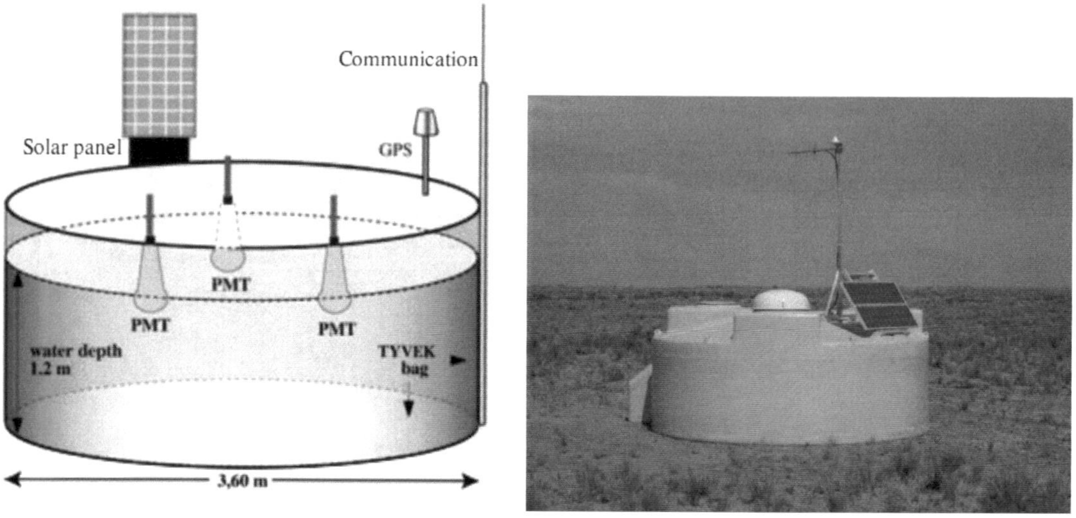

Figure 3. Schematic view of a water Cherenkov station (left) and photograph (right) of a Surface Detector unit.

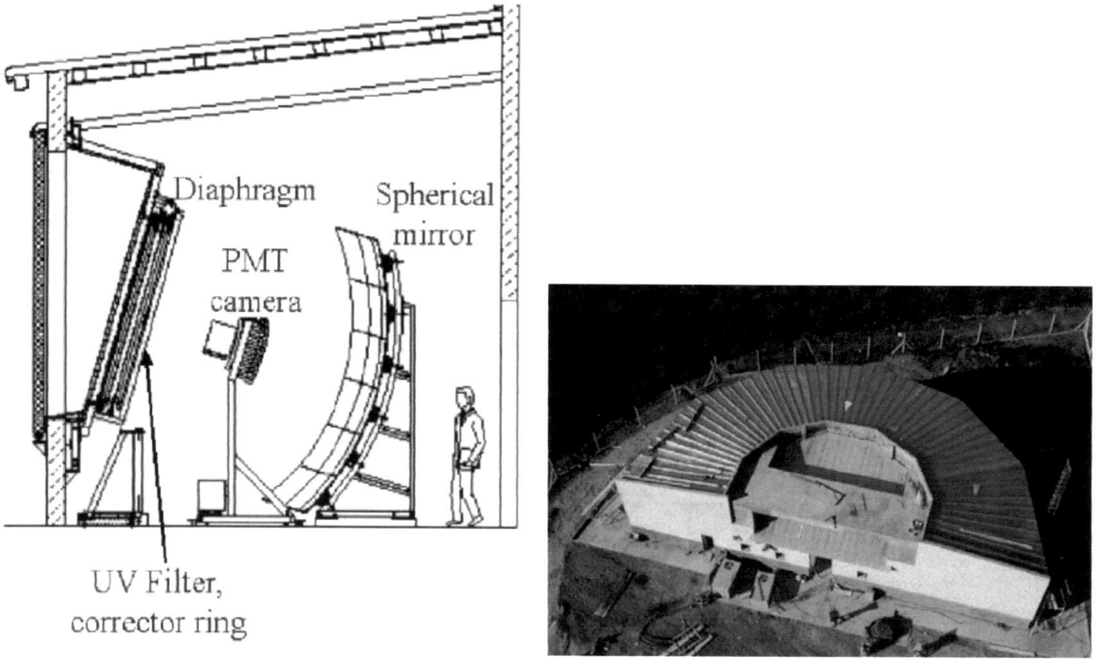

Figure 4. Schematic view of a fluorescence telescope (left) and aerial view (right) of the FD building at Los Leones.

3. Low-energy Enhancements

3.1. The AMIGA Infill extension

The low-energy extension of the surface detector of the observatory is the Auger Muon and Infill for the Ground Array (AMIGA [11, 12]). The Infill array consists of an area of 23.5 km^2 with additional water-Cherenkov stations on a denser grid. The additional stations are set out forming an hexagonal spacing with sides of 750 m, compared to the spacing of 1500 m of the main array. This infilled array is close to and in the field of view of HEAT. Since September 2011 all the 61 surface stations planned for the 750 m infill have been deployed. Data taking with the Infill array still under partial construction began in 2008.

The analysis of the data collected, based on the algorithms developed for the regular array, is in an advanced stage. A scheme of Infill array is shown in figure 5.

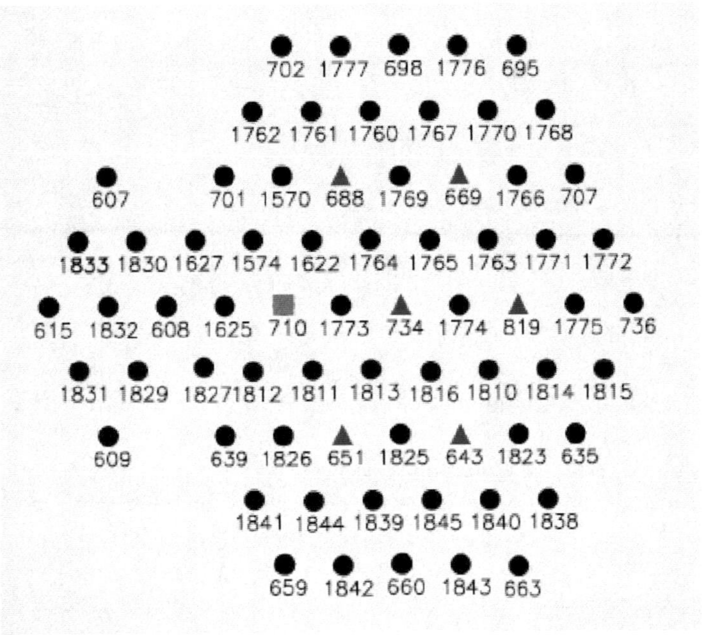

Figure 5. A scheme of the Infill array, HEATLET array and Coihueco and HEAT fluorescence telescopes. The additional Infill stations are marked in red and HEATLET stations in pink.

In addition to the Infill array, AMIGA consists of an associated set of muon detectors (MDs). The MDs are scintillator counters buried close to the 61 infill stations under 2.3 m of soil. Each station consists of two scintillator modules of 10 m^2 and two smaller modules of 5 m^2. Their muon energy threshold is about 1 GeV. The MD is currently in its prototype phase, named the Unitary Cell (UC). The UC will consist of 7 buried detectors to be installed on one hexagon and on its centre. And is foreseen for early 2012.

The water-Cherenkov detectors used in the Infill array are the same used in the main array. Moreover, the Infill is embedded in the regular array and therefore similar strategies may be employed for the selection and reconstruction of the observed showers. The trigger efficiency, the aperture and exposure calculation, the event selection, the geometry reconstruction and lateral distribution functions used (LDFs), as well as the energy estimator and the energy calibration, all benefit from algorithms tested successfully over the past years for the main array of the Auger SD.

The trigger system of the Infill array is adopted from the regular Auger array [8]. An event

Journal of Physics: Conference Series **468** (2013) 012013 doi:10.1088/1742-6596/468/1/012013

is accepted when at least 3 stations forming a triangle (see figure 6) satisfy a local trigger of the type Time-over-Threshold (3ToT event). The smaller spacing between stations of the Infill leads to an increase of the trigger efficiency at low energy, as illustrates figure 7 for zenith angles < 55°, for both Infill and regular array. As is shown, the 750 m spacing of the Infill allows cosmic rays to be detected with efficiency of 100% for energies above $3\text{x}10^{17}$ eV.

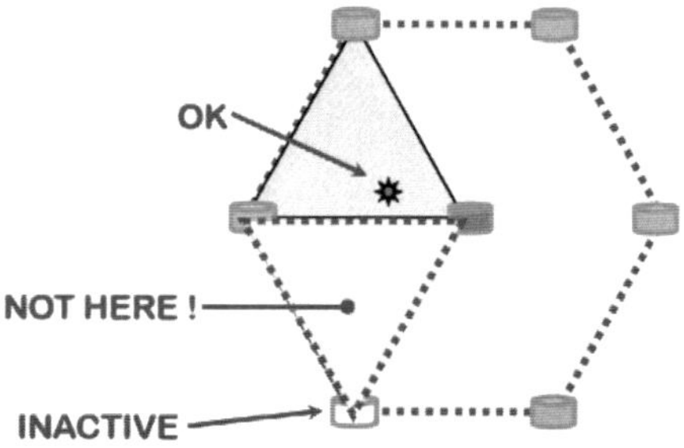

Figure 6. Event accepted when at least 3 stations forming a triangle satisfy a local trigger of the type Time-over-Threshold (3ToT event) .

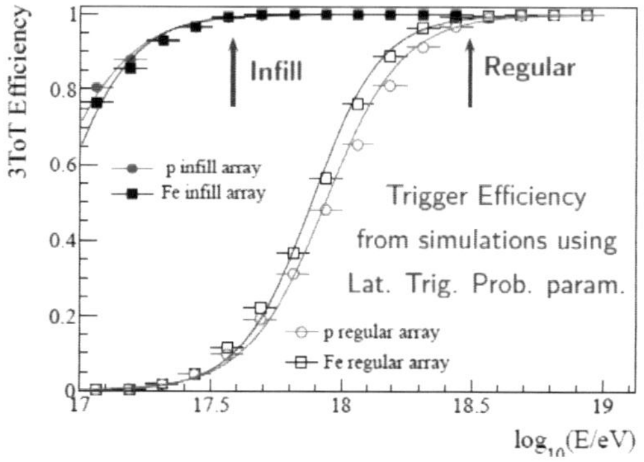

Figure 7. 3ToT trigger efficiency for the Infill and regular array obtained from simulations of Iron and proton primaries.

The 3ToT trigger rate is around 55 events/day/hexagon out of which around 28 events/day/hexagon satisfy the fiducial event selection that requires 6 working stations to surround the station with the highest signal, i.e. a working hexagon. This condition assures a good reconstruction and allows for a simple geometrical calculation of the aperture. Integrating the instantaneous effective area over the time when the detector was stable, the acceptance between August 2008 and March 2011 amounts to 26.4±1.3 km^2 sr yr.

As was done for the principal array (1500 m separation) measurement uncertainties have been derived from data. The angular resolution of the 750 m array was found to be 1.3° for events with at least 4 stations. To reconstruct events, the distribution of SD signals on ground as a function of the distances to the shower axis, S(r), is fitted with a Lateral Distribution Function LDF . The optimum distance, r_{opt}, where the signal fluctuations are minimised depends on the array spacing. This signal, $S(r_{opt})$, is the ground parameter eventually used to obtain an energy estimator. For the regular array $r_{opt} = 1000$ m whereas for the 750 m Infill was found to be 450 m. Besides the uniform treatment of data from the highest energies down to $3\text{x}10^{17}$ eV, an additional advantage of having the Infill within the regular array is that it is possible to make cross checks of results in the overlap region. From event to event, the main statistical sources of uncertainty in the parameter S(450) are the shower-to-shower fluctuations, the finite size of the detectors and the sparse sampling of the LDF. There is also the systematic contribution due to the lack of knowledge of the LDF. The total S(450) uncertainty derived from the Infill data goes from ~20% at 10 VEM (Vertical Equivalent Muon) to ~10% at 100 VEM. The energy of an event is estimated from the ground parameter accounting empirically for the shower attenuation in the atmosphere (that depends on the zenith angle) with a constant intensity cut method. The equivalent signal at a reference zenith angle of 35° (S_{35}) is used to infer the energy. S_{35} is calibrated using hybrid events, and correlates with the energy measured with the FD, as illustrated in figure 8. The number of events as a function of S_{35} is illustrated in figure 9 for different zenith angle intervals, and the Infill array was found to be fully efficient from S_{35} ~20 VEM, the trigger threshold.

Journal of Physics: Conference Series **468** (2013) 012013 doi:10.1088/1742-6596/468/1/012013

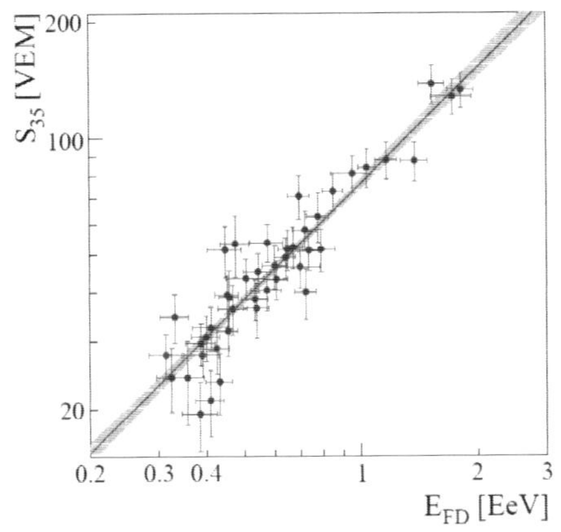

Figure 8. The correlation between S_{35} and energy.

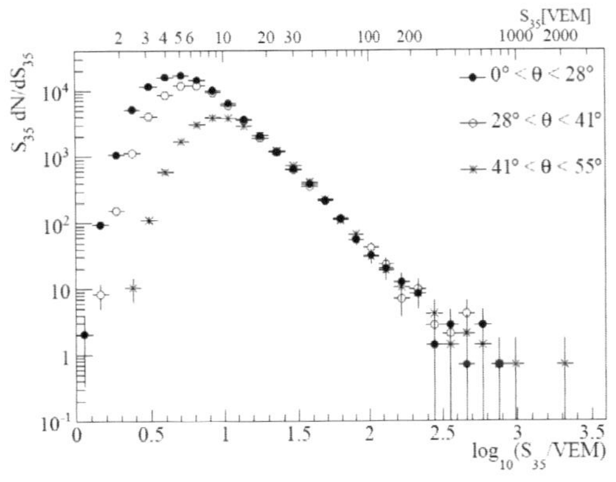

Figure 9. The distribution of events as a function of $\log_{10}(S_{35}/\mathrm{VEM})$ for different zenith angle intervals.

3.2. HEAT

The low-energy extension of the fluorescence detector of the Observatory are the 3 High Elevation Auger Telescopes (HEAT) [13] (figure 10), which operate on the same principles as the standard fluorescence telescopes of the Observatory.

Figure 10. Photo of the 3 HEAT telescopes tilted upward. In the background the telecommunication tower of Coihueco is visible.

HEAT telescopes are located near the FD Coihueco building, overlooking the Infill array and additional water-Cherenkov stations installed to further increase the effective Infill area for low energy shower measurements (HEAT Low Energy Trigger, HEATLET). The HEAT telescopes are very similar to the FD telescopes but are housed in three individual buildings, which can be tilted hydraulically by 30 degrees to cover the range from 30 to 60 degrees above horizon, so HEAT has two modes of operation: Down Mode (figure 11) and Up Mode (figure 12). The horizontal mode of the HEAT telescopes, which is used for installation, commissioning and maintenance of the hardware, is also the position in which the absolute calibration of the telescopes takes place. In this position the field of view of the HEAT telescopes overlaps with those of the Coihueco telescopes. This offers the possibility of doing special analyses of events recorded simultaneously. In addition, these events can be used to check the alignment of the new telescopes and provide a cross-check of their calibration constants.

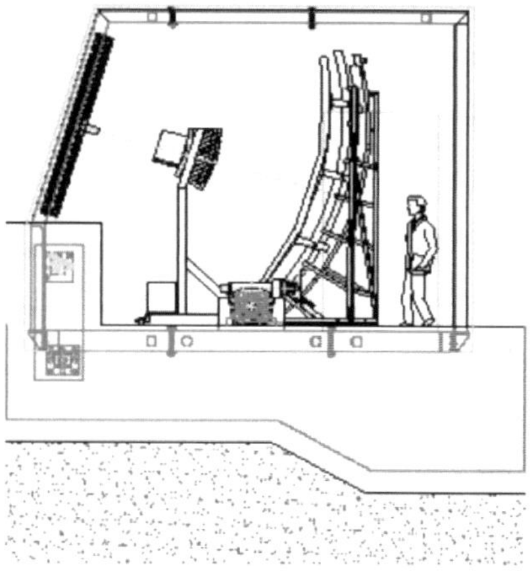

Figure 11. Down Mode: horizontal mode for service and cross-calibration.

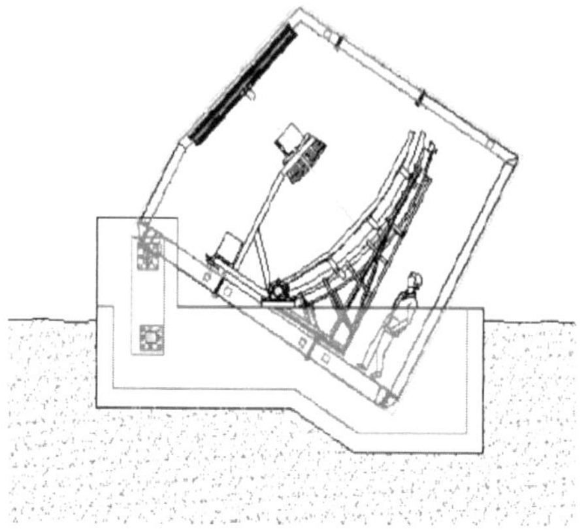

Figure 12. Up Mode: data-taking mode in tilted orientation.

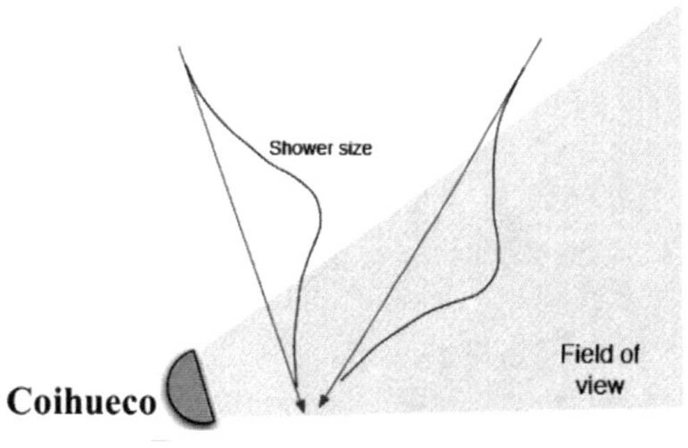

Figure 13. Limited field of view on reconstruction of standard FD: Showers approaching the telescope have much higher reconstruction probability than those departing.

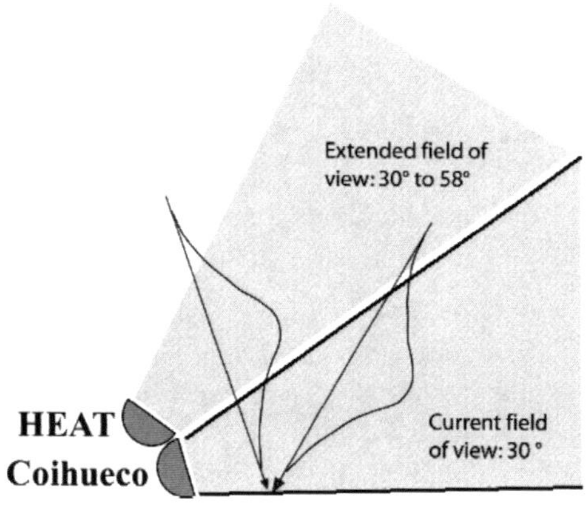

Figure 14. Extended field of view by combination of Coihueco+HEAT.

The fluorescence technique is best suited to determine the cosmic ray composition by a direct measurement of the shower depth maximum X_{max}. HEAT allows to lower the energy threshold for these measurements. As the fluorescence light signal is roughly proportional to the primary particle energy, low energy showers can be detected only at close distances. The field of view of the baseline design of the FD is limited to 30° above the horizon (see figure 13). At close distances only the lowest few kilometres of the atmosphere are in the field of view, but, low energy showers reach their maximum of development at higher altitudes. Thus, the crucial region around the shower maximum is generally not observable. The small fraction of the shower development, which falls within the field of view, is mostly insufficient to determine the depth of shower maximum X_{max}. In addition, this cut-off effect also depends on primary mass and shower direction. A plain reconstruction of the shower profile using raw data would yield biased results with respect to zenith angle and mass composition.

With HEAT in the tilted position (mode up), the combined HEAT-Coihueco telescopes cover an elevation range from the horizon to 60° (figure 14). This extended field of view enables the reconstruction of low energy showers for close-by shower events and resolves ambiguities in the X_{max} determination. The improved energy resolution and X_{max} determination is especially visible in the low energy regime.

An example of one of the low-energy showers recorded with HEAT and Coihueco is shown in figure 15 (Camera image of the recorded signal) and figure 16 (Reconstructed energy deposit profile). The energy of this event is $\sim 1.07 \times 10^{17}$ eV.

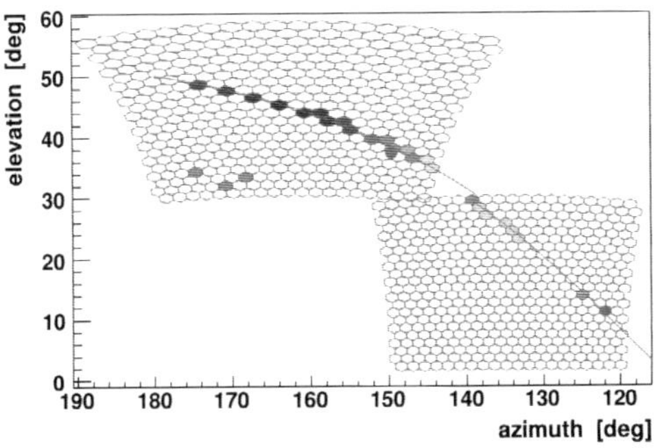

Figure 15. This shower was recorded by HEAT and Coihueco.The relative arrival time of the fluorescence light is coded in the colour of the pixel. The solid line is a fit of the shower detector plane.

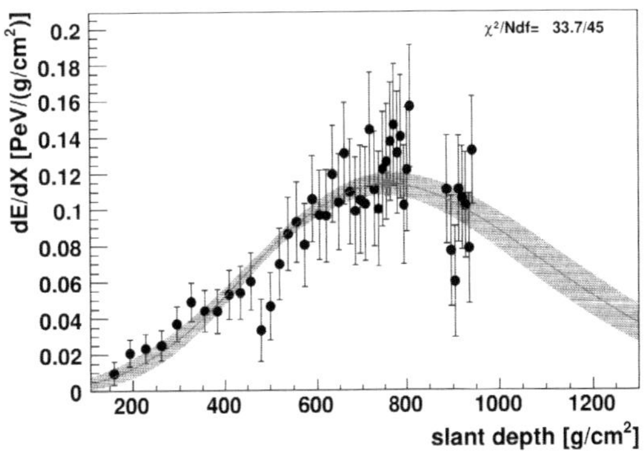

Figure 16. Longitudinal shower profile together with Gaisser-Hillas-fit. Only a fit using both HEAT (left) and Coihueco (right) data points results in a reasonable X_{max} value.

Since June 2010, the data taking and data quality reached a satisfactory performance level. The alignment of the regular fluorescence telescopes is obtained from star tracking. In addition to this method, a new one was introduced to determine the alignment of HEAT telescopes. Given a reference geometry from any number of sources (SD, hybrid, reconstruction from other sites, laser shots) and the observation of the corresponding light traces in a HEAT telescope, the developed algorithm determines the optimal pointing direction for the telescope. The accuracy of the method increases when applied to several events. This method results in a statistical accuracy of 0.3° or better for elevation and azimuth. For the HEAT telescope 1, which has the Central Laser Facility (CLF) in its field of view, accuracies of better than 0.1° can be achieved.

For shower reconstruction it is desired to combine the data from HEAT and Coihueco. However, in the standard reconstruction chain the data from each building is used separately. Thus, the official software (Offline) has been generalized. The different telescopes can then be combined in any desired form to build a virtual site.

3.2.1. HEATLET (Heat Low Energy Trigger tanks), consists of 5 additional water-Cherenkov stations forming an array of 9 Infill stations (see figure 17), very close to HEAT, in order to get the best low-energy hybrid event reconstructions. The timing information from at least one triggered surface station in a hybrid event enables showers at low energies to be reconstructed with good accuracy. The single station trigger probability needed for such hybrid measurements depends on the detector spacing, as well as on the energy, zenith angle, and mass of the primary particle. In figure 18 an example of an event as described above and recorded with the HEATLET array is displayed.

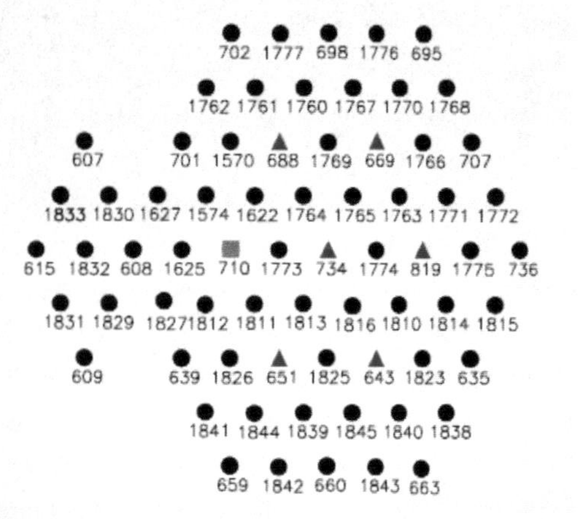

Figure 17. Layout of HEATLET and Infill

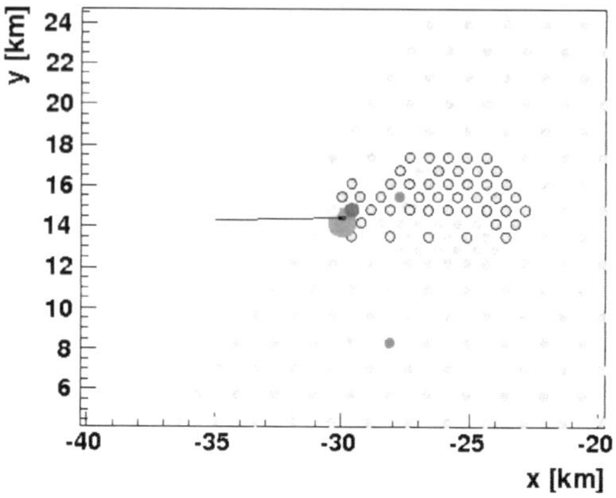

Figure 18. Event in HEATLET array.

4. Conclusions

The Infill array, part of the AMIGA enhancement of the Pierre Auger Observatory, has been operating in good conditions since its deployment. It extends the energy range for the surface detector of Pierre Auger Observatory down to 3x10^{17} eV. The integrated exposure between August 2008 and March 2011 is (26.4 $\pm$1.3) km^2 sr yr, with resolution in S(450) varying from 20% at 10 VEM to 10% at 100 VEM, at the highest energies.

The HEAT telescopes have operated since September 2009 and are producing high quality data in a stable manner since June 2010. These new telescopes improve significantly the quality of data for energy and mass composition analyses at low energies. The enhancements to the baseline of the Pierre Auger Observatory also include, in addition to AMIGA and HEAT, arrays for detection of radio and GHz emission from air showers [14, 15, 16]

References

[1] Linsley J 1963 *Phys. Rev. Lett.* **10** 146
[2] Bird D J et al 1993 Evidence for correlated changes in the spectrum and composition of cosmic rays at extremely high energies *Phys. Rev. Lett.* **71** 3401
Bird D J et al 1994 The cosmic-ray energy spectrum observed by the fly's eye *Astrophys. J.* **424** 491
[3] Blümer J et al 2009 *Nucl. Phys.* **63** 293
[4] Meurer C, Scharf N, for the Pierre Auger Collaboration 2011 Heat a low energy enhancement of the Pierre Auger Observatory *Astrophys Space Sci. Trans.* **7** 183-6
[5] Abraham J et al 2004 *Nucl. Instr. and Meth.* A **523** 50
[6] Allekotte I et al 2008 *Nucl. Instr. Meth.* A **586** 409
[7] Pierre Auger Collaboration 2010 *Nucl. Instr. Meth.* A **620** 227
[8] Pierre Auger Collaboration 2010 *Nucl. Instr. Meth.* A **613** 29-39
[9] Berezinsky V, Gazizov A Z and Grigorieva S I 2005 *Phys. Lett.* B **612** 147
[10] Aloisio R, Berezinsky V, Blasi P and Ostapchenko S 2008 *Phys. Rev.* D**77**
[11] Sánchez F, for the Pierre Auger Collaboration 2011 *Proc. 32nd Int. Cosmic Ray Conf.* (Beijing, China) arXiv:1107.4807
[12] Maris I C, for the Pierre Auger Collaboration 2011 *Proc. 32nd Int. Cosmic Ray Conf.* (Beijing, China) arXiv:1107.4809
[13] Mathes T H-J, for the Pierre Auger Collaboration 2011 *Proc. 32nd Int. Cosmic Ray Conf.* (China, Beijing) arXiv:1107.4807
[14] Kelley J, for the Pierre Auger Collaboration 2011 *Proc. 32nd Int. Cosmic Ray Conf.* (China, Beijing) arXiv:1107.4807
[15] Revenu B, for the Pierre Auger Collaboration 2011 *Proc. 32nd Int. Cosmic Ray Conf.* (China, Beijing) arXiv:1107.4807

[16] Allison P, for the Pierre Auger Collaboration 2011 *Proc. 32nd Int. Cosmic Ray Conf.* (China, Beijing) arXiv:1107.4807

Journal of Physics: Conference Series **468** (2013) 012014 doi:10.1088/1742-6596/468/1/012014

Renormalization of the baryon axial vector current in large-N_c chiral perturbation theory

Rubén Flores-Mendieta[1], **María A. Hernández Ruíz**[2], **and Christoph P. Hofmann**[3]

[1] Instituto de Física, Universidad Autónoma de San Luis Potosí,
Álvaro Obregón 64, Zona Centro, San Luis Potosí, San Luis Potosi 78000, México.
[2]Facultad de Ciencias Químicas, Universidad Autónoma de Zacatecas,
Apartado Postal 585, Zacatecas, Zacatecas 98060, México.
[3] Facultad de Ciencias, Universidad de Colima,
Bernal Díaz del Castillo 340, Colima, Colima 28045, México.

E-mail: `mahernan@uaz.edu.mx`

Abstract. The baryon axial vector current is considered within the combined framework of large-N_c baryon chiral perturbation theory (where N_c is the number of colors) and the baryon axial vector couplings are extracted. Loop graphs with octet and decuplet intermediate states are systematically incorporated into the analysis.

1. Introduction

The theory of the strong interactions is quantum chromodynamics (QCD). Different methods have been used to extract low-energy consequences of QCD. In this work, we use a combined expansion in m_q (where m_q is the quark mass) and $1/N_c$ [1]. The $1/N_c$ chiral effective Lagrangian for the lowest-lying baryons was constructed in Ref. [2].

On the one hand, chiral perturbation theory exploits the symmetry of the QCD Lagrangian under $SU(3)_L \times SU(3)_R \times U(1)_V$ transformations on the three flavors of light quarks u, d and s in the limit that the quark masses m_u, m_d and m_s vanish. Chiral symmetry is spontaneously broken to the vector subgroup $SU(3) \times U(1)_V$ by the QCD vacuum, giving rise to the octet of pseudoscalar Goldstone bosons (π, K and η). When chiral perturbation theory is extended to include baryons, it is convenient to introduce velocity-dependent baryon fields [3], so that the expansion of the baryon chiral Lagrangian in powers of m_q and $1/M_B$ (where M_B is the baryon mass) is manifest. This is the so-called heavy baryon chiral perturbation theory [3]. The inclusion of decuplet baryon intermediate states yields sizable cancellations between one loop corrections [4]. This phenomenological observation can be explained in the context of the $1/N_c$ expansion.

On the other hand, large-N_c QCD is the $SU(N_c)$ gauge theory of quarks and gluons where the number of colors, N_c, is a parameter of the theory [2]. Large-N_c is the generalization of QCD from $N_c = 3$ to $N_c \gg 3$ colors. A spin-flavor symmetry emerges for baryons in the large-N_c limit and can be used to classify large-N_c baryon states and matrix elements [2], which has led to remarkable insights into the understanding of the nonperturbative QCD dynamics of hadrons.

In particular, in this work we will describe the baryon axial-vector couplings, and as a result we obtain corrections at relative orders $1/N_c$ and $1/N_c^2$.

2. The chiral Lagrangian for baryons in the $1/N_c$ expansion

The $1/N_c$ chiral Lagrangian for baryons reads [2]

$$\mathcal{L}_{\text{baryon}} = i\mathcal{D}^0 - \mathcal{M}_h + Tr(\mathcal{A}^k\lambda^c)A^{kc}\frac{1}{N_c}Tr\left(\mathcal{A}^k\frac{2I}{\sqrt{6}}\right)A^k + \ldots \tag{1}$$

where

$$\mathcal{D}^0 = \partial^0 1 + Tr(\mathcal{V}^0\lambda^c)T^c. \tag{2}$$

Each term in Eq. (1) involves a baryon operator which can be expressed as a polynomial in the SU(6) spin-flavor generators [2]

$$J^k = q^\dagger\frac{\sigma^k}{2}q, \qquad T^c = q^\dagger\frac{\lambda^c}{2}q, \qquad G^{kc} = q^\dagger\frac{\sigma^i}{2}\frac{\lambda^a}{2}q, \tag{3}$$

where $q^\dagger$ and q are $SU(6)$ operators that create and annihilate states in the fundamental representation of $SU(6)$, and σ^k and λ^c are the Pauli spin and Gell-Mann flavor matrices, respectively.

The baryon operator $\mathcal{M}_h$ denotes the spin splittings of the tower of baryon states with spins $1/2, \ldots, N_c/2$ in the flavor representations. Furthermore, the vector and axial vector combinations of the meson fields,

$$\mathcal{V}^0 = \frac{1}{2}(\xi\partial^0\xi^\dagger + \xi^\dagger\partial^0\xi), \qquad \mathcal{A}^k = \frac{i}{2}(\xi\nabla^k\xi^\dagger - \xi^\dagger\nabla^k, \xi),$$

couple to baryon vector and axial vector currents, respectively. Here $\xi = \exp[i\Pi(x)/f]$, where $\Pi(x)$ stands for the nonet of Goldstone boson fields and $f \approx 93$ MeV is the meson decay constant.

The QCD operators involved in $\mathcal{L}_{baryon}$ in Eq. (1) have well-defined $1/N_c$ expansions. Specifically, the baryon axial vector current A^{kc} is a spin-1 object, an octet under $SU(3)$, and odd under time reversal. Its $1/N_c$ expansion reads

$$A^{kc} = a_1 G^{kc} + \sum_{n=2,3}^{N_c} b_n\frac{1}{N_c^{n-1}}\mathcal{D}_n^{kc} + \sum_{n=3,5}^{N_c} c_n\frac{1}{N_c^{n-1}}\mathcal{O}_n^{kc}, \tag{4}$$

where the unknown coefficients a_1, b_n, and c_n have expansions in powers of $1/N_c$ and are order unity at leading order in the $1/N_c$ expansion. The first few operators in expansion (4) are

$$\mathcal{D}_2^{kc} = J^k T^c, \tag{5}$$

$$\mathcal{D}_3^{kc} = \{J^k, \{J^r, G^{rc}\}\}, \tag{6}$$

$$\mathcal{O}_3^{kc} = \{J^2, G^{kc}\} - \frac{1}{2}\{J^k, \{J^r, G^{rc}\}\}, \tag{7}$$

while higher order terms can be obtained as $\mathcal{D}_n^{kc} = \{J^2, \mathcal{D}_{n-2}^{kc}\}$ and $\mathcal{O}_n^{kc} = \{J^2, \mathcal{O}_{n-2}^{kc}\}$ for $n \geq 4$. Notice that $\mathcal{D}_n^{kc}$ are diagonal operators with non-zero matrix elements only between states with the same spin, and the $\mathcal{O}_n^{kc}$ are purely off-diagonal operators with non-zero matrix elements only between states with different spin. At $N_c = 3$ the series (4) can be truncated as

$$A^{kc} = a_1 G^{kc} + b_2\frac{1}{N_c}\mathcal{D}_2^{kc} + b_3\frac{1}{N_c^2}\mathcal{D}_3^{kc} + c_3\frac{1}{N_c^2}\mathcal{O}_3^{kc}. \tag{8}$$

The matrix elements of the space components of A^{kc} between SU(6) symmetric states yield the values of the axial vector couplings. For the octet baryons, the axial vector couplings are g_A, as defined in experiments in baryon semileptonic decays.

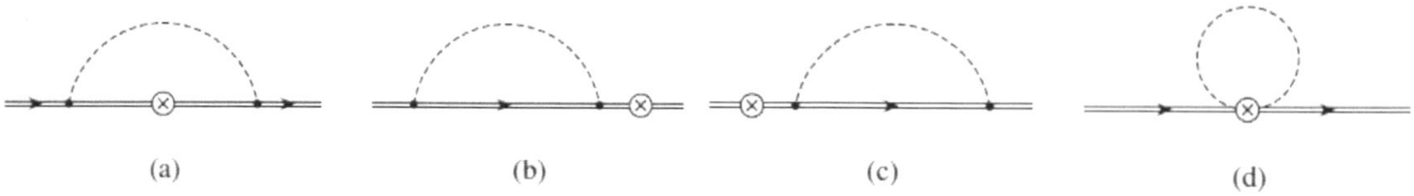

Figure 1. One-loop corrections to the baryon axial vector current.

3. Renormalization of the baryon axial vector current

The baryon axial vector current A^{kc} is renormalized by the one-loop diagrams displayed in Fig. 1. These loop graphs have a calculable dependence on the ratio Δ/m_Π, where $\Delta \equiv M_\Delta - M_N$ is the decuplet-octet mass difference and m_Π is the meson mass.

The correction arising from the sum of the diagrams of Figs. 1(a)-1(c), containing the full dependence on the ratio Δ/m_Π, reads [4]

$$\begin{aligned}\delta A^{kc} &= \frac{1}{2}\left[A^{ja},\left[A^{jb},A^{kc}\right]\right]\Pi^{ab}_{(1)} - -\frac{1}{2}\{A^{ja},\left[A^{kc},\left[\mathcal{M},A^{jb}\right]\right]\}\Pi^{ab}_{(2)} \\ &+ \frac{1}{6}\left(\left[A^{ja},\left[\left[\mathcal{M},\left[\mathcal{M},A^{jb}\right]\right],A^{kc}\right]\right] - \frac{1}{2}\left[\left[\mathcal{M},A^{ja}\right],\left[\left[\mathcal{M},A^{jb}\right],A^{kc}\right]\right]\right)\Pi^{ab}_{(3)} + \ldots\end{aligned}$$

Here $\Pi^{ab}_{(n)}$ is a symmetric tensor which contains meson-loop integrals with the exchange of a single meson: A meson of flavor a is emitted and a meson of flavor b is reabsorbed. $\Pi^{ab}_{(n)}$ descomposes into flavor singlet, flavor **8** and flavor **27** representations

$$\Pi^{ab}_{(n)} = F^{(n)}_{\mathbf{1}}\delta^{ab} + F^{(n)}_{\mathbf{8}}d^{ab8} + F^{(n)}_{\mathbf{27}}\left[\delta^{a8}\delta^{b8} - \frac{1}{8}\delta^{ab} - \frac{3}{5}d^{ab8}d^{888}\right], \tag{9}$$

where

$$\begin{aligned}F^{(n)}_{\mathbf{1}} &= \frac{1}{8}\left[3F^{(n)}(m_\pi,0,\mu) + 4F^{(n)}(m_K,0,\mu) + F^{(n)}(m_\eta,0,\mu)\right], \\ F^{(n)}_{\mathbf{8}} &= \frac{2\sqrt{3}}{5}\left[\frac{3}{2}F^{(n)}(m_\pi,0,\mu) - F^{(n)}(m_K,0,\mu) - \frac{1}{2}F^{(n)}(m_\eta,0,\mu)\right], \\ F^{(n)}_{\mathbf{27}} &= \frac{1}{3}F^{(n)}(m_\pi,0,\mu) - \frac{4}{3}F^{(n)}(m_K,0,\mu) + F^{(n)}(m_\eta,0,\mu).\end{aligned}$$

Explicit expressions for the general function $F^{(n)}(m_\Pi,\Delta,\mu)$, defined by

$$F^{(n)}(m_\Pi,\Delta,\mu) \equiv \frac{\partial^n F(m_\Pi,\Delta,\mu)}{\partial\delta^n}, \tag{10}$$

can be found in Ref. [5]

4. Results and Conclusions

The analysis was performed at one-loop order, where the corrections to the baryon axial vector coupling arise at relative orders $1/N_c$, $1/N_c^2$, and so on, which is precisely the origin of the $1/N_c$ expansion. The predicted values for g_A are listed in Table 1. Our final results referring to the degeneracy limit have been analyzed in Ref. [1, 5].

Table 1. Relative orders $1/N_c$ to the coupling constants g_A.

			Figs. 1(a-d)					
			1		**8**		**27**	
Process	Total	Tree	$\mathcal{O}(\frac{1}{N_c})$	$\mathcal{O}(\frac{1}{N_c^2})$	$\mathcal{O}(\frac{1}{N_c})$	$\mathcal{O}(\frac{1}{N_c^2})$	$\mathcal{O}(\frac{1}{N_c})$	$\mathcal{O}(\frac{1}{N_c^2})$
$n \to pe^-\bar{\nu}_e$	1.275	1.238	0.480	−0.549	−0.181	0.278	−0.001	0.009
$\Sigma^\pm \to \Lambda e^+\nu_e$	0.623	0.661	0.279	−0.319	−0.040	0.047	−0.004	0
$\Lambda \to pe^-\bar{\nu}_e$	−0.899	−0.855	−0.317	0.360	−0.007	−0.089	0.005	0.005
$\Sigma^- \to ne^-\bar{\nu}_e$	0.345	0.381	0.457	−0.488	−0.005	−0.002	−0.002	0.004
$\Xi^- \to \Lambda e^-\bar{\nu}_e$	0.225	0.194	0.062	−0.064	0.032	0.010	−0.007	−0.002
$\Xi^- \to \Sigma^0 e^-\bar{\nu}_e$	0.795	0.875	0.338	−0.387	0.064	−0.098	−0.006	0.008
$\Xi^0 \to \Sigma^+ e^-\bar{\nu}_e$	1.124	1.238	0.480	−0.549	0.091	−0.139	0.013	0.020

Table 1 shows the numerical values of the g_A axial vector couplings for various semileptonic processes in the $1/N_c$ expansion, individually for the flavor singlet **1**, octet **8**, and **27** contributions. The singlet corrections are $1/N_c$ suppressed with respect to the tree-level value. Subsequent suppressions of the octet and **27** contributions are also noticeable. The results are perfectly consistent both with the expectations from the $1/N_c$ expansion and the experimental data.

Acknowledgments

We acknowledge support from CONACyT-SNI and PROMEP (México).

References

[1] R. Flores-Mendieta, M. A. Hernndez-Ruiz, C. P. Hofmann, Phys. Rev. D **86**, 094041 (2012).
[2] E. Jenkins, Phys. Rev. D **53**, 2625 (1996).
[3] E. Jenkins and A. V. Manohar, Phys. Lett. B **255** 558 (1991) and E. Jenkins and A. V. Manohar, Phys. Lett. B **259** 353 (1991).
[4] R. Flores-Mendieta, C. P. Hofmann, E. E. Jenkins, Phys. Rev. D **61**, 116014 (2000).
[5] R. Flores-Mendieta, C. P. Hofmann, Phys. Rev. D **74**, 094001 (2006).

AUTHOR INDEX

Institute of Physics
Dirac House, Temple Back
Bristol BS1 6BE UK

proceedings
.com